Helmets and Body Armour in
New Kingdom Egypt

Bloomsbury Egyptology

Ancient Egyptians at Play, Walter Crist, Anne-Elizabeth Dunn-Vaturi and Alex de Voogt
Ancient Egyptian Technology and Innovation, Ian Shaw
Archaeologists, Tourists, Interpreters, Rachel Mairs and Maya Muratov
Asiatics in Middle Kingdom Egypt, Phyllis Saretta
Burial Customs in Ancient Egypt, Wolfram Grajetzki
Court Officials of the Egyptian Middle Kingdom, Wolfram Grajetzki
The Egyptian Oracle Project, edited by Robyn Gillam and Jeffrey Jacobson
Foreigners in Ancient Egypt, Flora Brooke Anthony
Five Egyptian Goddesses: Their Possible Beginnings, Actions, and Relationships in the Third Millennium BCE, Susan Tower Hollis
Hidden Hands, Stephen Quirke
Subsistence Strategies and Craft Production at the Ancient Egyptian Ramesside Fort of Zawiyet Umm el-Rakham, Nicky Nielsen
The Middle Kingdom of Ancient Egypt, Wolfram Grajetzki
The Unknown Tutankhamun, Marianne Eaton-Krauss
Performance and Drama in Ancient Egypt, Robyn Gillam

Helmets and Body Armour in New Kingdom Egypt

Alberto Maria Pollastrini

BLOOMSBURY ACADEMIC
LONDON • NEW YORK • OXFORD • NEW DELHI • SYDNEY

BLOOMSBURY ACADEMIC

Bloomsbury Publishing Plc, 50 Bedford Square, London, WC1B 3DP, UK
Bloomsbury Publishing Inc, 1359 Broadway, 12th Floor, New York, NY 10018, USA
Bloomsbury Publishing Ireland, 29 Earlsfort Terrace, Dublin 2, D02 AY28, Ireland

BLOOMSBURY, BLOOMSBURY ACADEMIC and the Diana logo are
trademarks of Bloomsbury Publishing Plc

First published in Great Britain 2024
This paperback edition published 2026

Copyright © Alberto Maria Pollastrini, 2024

Alberto Maria Pollastrini has asserted his right under the Copyright, Designs and
Patents Act, 1988, to be identified as Author of this work.

For legal purposes the Acknowledgements on p. xiv constitute an
extension of this copyright page.

Cover design: Terry Woodley
Cover image: Paintings copied from the tomb of Ramses IV.
© The Nature Notes/Alamy Stock Photo

All rights reserved. No part of this publication may be: i) reproduced or transmitted in
any form, electronic or mechanical, including photocopying, recording or by means of any
information storage or retrieval system without prior permission in writing from the publishers;
or ii) used or reproduced in any way for the training, development or operation of artificial intelligence
(AI) technologies, including generative AI technologies. The rights holders expressly reserve this
publication from the text and data mining exception as per Article 4(3) of the Digital Single
Market Directive (EU) 2019/790.

Bloomsbury Publishing Inc does not have any control over, or responsibility for,
any third-party websites referred to or in this book. All internet addresses given
in this book were correct at the time of going to press. The author and publisher
regret any inconvenience caused if addresses have changed or sites have
ceased to exist, but can accept no responsibility for any such changes.

A catalogue record for this book is available from the British Library.

Library of Congress Cataloging-in-Publication Data
Names: Pollastrini, Alberto Maria, author.
Title: Helmets and body armour in New Kingdom Egypt / Alberto Maria Pollastrini.
Other titles: Bloomsbury Egyptology; v. 11.
Description: New York: Bloomsbury Academic, 2024. | Series: Bloomsbury Egyptology;
vol 11 | Includes bibliographical references and index.
Identifiers: LCCN 2023055800 (print) | LCCN 2023055801 (ebook) |
ISBN 9781350323483 (hardback) | ISBN 9781350323490 (paperback) |
ISBN 9781350323506 (pdf) | ISBN 9781350323513 (ebook)
Subjects: LCSH: Armor, Ancient–Egypt. | Helmets–Egypt. | Egypt–History, Military.
Classification: LCC U805 .P65 2024 (print) | LCC U805 (ebook) |
DDC 355.8/2418093209013–dc23/eng/20231229
LC record available at https://lccn.loc.gov/2023055800
LC ebook record available at https://lccn.loc.gov/2023055801

ISBN: HB: 978-1-3503-2348-3
PB: 978-1-3503-2349-0
ePDF: 978-1-3503-2350-6
eBook: 978-1-3503-2351-3

Series: Bloomsbury Egyptology

Typeset by RefineCatch Limited, Bungay, Suffolk

For product safety related questions contact productsafety@bloomsbury.com.

To find out more about our authors and books visit www.bloomsbury.com
and sign up for our newsletters.

I am tired and sick of war. Its glory is all moonshine. It is only those who have neither fired a shot nor heard the shrieks and groans of the wounded who cry aloud for blood, for vengeance, for desolation.

War is hell.

William Tecumseh Sherman

To Ilaria, Morgana and Ginevra

Contents

List of Illustrations	x
Preface and Acknowledgements	xiv
Note for the Reader	xvi
List of Abbreviations	xvii
1 Introduction	1
2 Reception and Diffusion of Personal Protections during the Eighteenth Dynasty	21
3 Personal Protective Equipment during the Ramesside Period	45
4 The Manufacture of Protective Gear in New Kingdom Egypt	63
5 Egyptian Terminology Relating to Protective Gear	89
6 Conclusions	103
Appendix I: Helmets as Depicted in Egyptian Art during the Eighteenth Dynasty	107
Appendix II: Helmets as Depicted in Egyptian Art during the Ramesside Period	109
Appendix III: Corslets as Depicted in Egyptian Art during the New Kingdom	111
Notes	113
Bibliography	155
Index	187

Illustrations

1.1 Military standards, weapons and other objects from the tomb of Ramesses III (KV 11). *Description de l'Égypte: Antiquités, Planches.* Tome II, pl. 88   4

1.2 *In-situ* drawing of helmets on crushed skulls from PG 789 tomb in the Royal Cemetery at Ur. Woolley, C. L., *The Royal Cemetery*, pl. 218 *a-b-c*   10

1.3 Reconstruction drawing of the copper alloy helmet from Tello. Cros, G., *RAAO*, Vol. 6, No. 3 (1906), 89   11

1.4 Copper plates found in the tomb of Mesehti, Asyut. Drawing by A. M. Pollastrini. (Not to scale)   16

2.1 Relief fragment ATP 09069 from the mortuary temple of Ahmose I at Abydos, possibly depicting a scale corslet. Drawing by A. M. Pollastrini   23

2.2 Right side of the chariot of Thutmose IV (CGC 46097). Carter, H. and Newberry, P. E., *The Tomb of Thoutmôsis IV*, pl. X   26

2.3 Left side of the chariot of Thutmose IV (CGC 46097). Carter, H. and Newberry, P. E., *The Tomb of Thoutmôsis IV*, pl. XI   26

2.4 Relief fragment from the mortuary temple of Horemheb reused in the forecourt of the temple of Khonsu at Karnak. The Epigraphic Survey, *Temple of Khonsu*. Vol. I, pl. 61. (Courtesy of the ISAC, Chicago)   30

2.5 Relief fragment from the mortuary temple of Horemheb reused in the forecourt of the temple of Khonsu at Karnak. Prisse d'Avennes, A., *Histoire de l'art égyptien*, pl. III. 13   31

2.6 Seti I subdues the town of Yenoam and Lebanon. Precint of Amun-Ra at Karnak, Hypostyle Hall, north wall, outer face. (Detail.) The Epigraphic Survey, *Relief and Inscriptions at Karnak*, Vol. IV, pl. 11. (Courtesy of the ISAC, Chicago)   32

2.7 Seti I subdues the town of Kadesh and the land of Amurru. Precinct of Amun-Ra at Karnak, Hypostyle Hall, north wall,

	Illustrations	xi
	outer face. The Epigraphic Survey, *Relief and Inscriptions at Karnak*, Vol. IV, pl. 23. (Courtesy of the ISAC, Chicago)	33
2.8	Hittite warriors killed by Ramesses II during the battle of Kadesh. Great Temple of Abu Simbel, Hypostyle Hall Drawing, north wall. Drawing by A. M. Pollastrini	33
2.9	Statuette of Western Asiatic (?) kneeling captive. Basel Museum of Ancient Art (Inv.-Nr. LgAe NN65). Di Natale, A. and Basile, C., eds, *Atti del XVIII Convegno di Egittologia e Papirologia, Siracusa, 20–23 Settembre 2018*, 149. (Courtesy of Museo del Papiro 'Corrado Basile', Siracusa)	34
3.1	The war council chaired by Ramesses II before the battle of Kadesh. Great Temple of Abu Simbel, Hypostyle Hall, north wall. (Detail.) Champollion, J.-F., *Monuments de l'Égypte et de la Nubie: Planches*. Tome I, pl. XXIX	46
3.2	Ramesses III distributes weapons to his soldiers before their campaign against the Sea Peoples. Mortuary Temple of Ramesses III at Medinet Habu, exterior wall, outer face, north side. (Detail.) The Epigraphic Survey, *Medinet Habu*, Vol. I, pl. 29. (Courtesy of the ISAC, Chicago)	48
3.3	The land battle of Djahy between Ramesses III and the Sea Peoples. Mortuary Temple of Ramesses III at Medinet Habu, exterior wall, outer face, north side. (Detail.) The Epigraphic Survey, *Medinet Habu*, Vol. I, pl. 33, 34. (Courtesy of the ISAC, Chicago)	50
3.4	The naval battle between the Egyptian fleet and the Sea Peoples. Mortuary Temple of Ramesses III at Medinet Habu, exterior wall, outer face, north side. (Detail.) The Epigraphic Survey, *Medinet Habu*, Vol. I, pl. 38, 39. (Courtesy of the ISAC, Chicago)	50
3.5	Ramesses III pursues the fleeing Libyans during the Second Libyan War (Year 11). Mortuary Temple of Ramesses III at Medinet Habu, exterior wall, outer face, north side, between pylons. (Detail.) The Epigraphic Survey, *Medinet Habu*, Vol. II, pl. 70. (Courtesy of the ISAC, Chicago)	51
3.6	Ramesses III defeats the Libyans during the Second Libyan War (Year 11). Mortuary Temple of Ramesses III at Medinet Habu,	

	first court, east wall. (Detail.) The Epigraphic Survey, *Medinet Habu*, Vol. II, pl. 72	51
3.7	Ramesses III storms a Syrian town. Mortuary Temple of Ramesses III at Medinet Habu, exterior wall, outer face, north side, between pylons. (Detail.) The Epigraphic Survey, *Medinet Habu*, Vol. II, pl. 90. (Courtesy of the ISAC, Chicago)	52
3.8	A Syrian town besieged by Ramesses III. Mortuary Temple of Ramesses III at Medinet Habu, first court, north side. (Detail.) The Epigraphic Survey, *Medinet Habu*, Vol. II, pl. 94. (Courtesy of the ISAC, Chicago)	53
3.9	The town of Tunip besieged by Ramesses III. Mortuary Temple of Ramesses III at Medinet Habu, exterior wall, outer face, north side, between pylons. (Detail.) The Epigraphic Survey, *Medinet Habu*, Vol. II, pl. 88. (Courtesy of the ISAC, Chicago)	53
3.10	Ramesses II on his chariot at the battle of Kadesh, wearing his scale corslet. Ramesseum, Second pylon, inner face. Prisse d'Avennes, A., *Histoire de l'art égyptien*, pl. III. 30	59
3.11	Standards and weapons. Tomb of Ramesses III (KV 11), Room M (side chamber Ch), left wall. Champollion, J.-F., *Monuments de l'Égypte et de la Nubie: Planches*. Tome III, pl. CCLXII	61
3.12	Standards and weapons. Tomb of Ramesses III (KV 11), Room M (side chamber Ch), right wall. Champollion, J.-F., *Monuments de l'Égypte et de la Nubie: Planches*. Tome III, pl. CCLXIII	62
4.1	Drawing reconstructing a portion of Ky-iry's tomb decoration. Grajetzki, W., *JEOL* 37, 115. (Coutesy of *Journal jaarbericht van het Vooraziatisch – Egyptische Genootschap (Gezelschap) 'Ex Oriente Lux'*)	64
4.2	Fragment of a Nineteenth Dynasty stele kept in the Museum of Fine Arts, Boston (MFA Accession n 1982.201). Drawing by A. M. Pollastrini	67
4.3	'Armour maker/armourer', Papyrus *Golenischeff* (Museum Pushkin 169). Drawing by A. M. Pollastrini	68
4.4	Stele of Nefer-renpet from Buhen (Penn Museum E10996). Drawing by A. M. Pollastrini	69

4.5	Copper alloy armour scales from Malqata (MMA 11.215.452 a–d.) Drawing by A. M. Pollastrini. (Not to scale)	71
4.6	Copper alloy armour scales from Malqata (MMA 11.215.452 e–h). Drawing by A. M. Pollastrini. (Not to scale)	72
4.7	Copper alloy armour scales from Malqata (MMA 11.215.452 i–j). Drawing by A. M. Pollastrini. (Not to scale)	73
4.8	Type 2 copper alloy scale from Šilwa-Tešub's house, Nuzi. Drawing by A. M. Pollastrini. (Not to scale)	74
4.9	Type 6 copper alloy scale from Šilwa-Tešub's house, Nuzi. Drawing by A. M. Pollastrini. (Not to scale)	75
4.10	Type 1 copper alloy scale from Kamidi (Kāmid el-Lōz). Drawing by A. M. Pollastrini. (Not to scale)	76
4.11	Copper alloy armour scales from Lisht (MMA 34.1.72, MMA 34.1.73). Drawing by A. M. Pollastrini. (Not to scale)	76
4.12	Set of scales made of different materials (copper alloy, bone, glazed ceramic and boar tusk ivory) from Qantir, area Q I, stratum B/2. Drawing by A. M. Pollastrini. (Not to scale)	79
4.13	Section of the pictorial papyrus British Museum EA 74100 showing Egyptian and Aegean infantrymen. Drawing by A. M. Pollastrini	81
4.14	Copper alloy armour scales 25.11.318 and 27.11.387 from Beth Shean. Drawing by A. M. Pollastrini. (Not to scale)	82
4.15	Copper alloy armour scales 25.11.501 and 27.11.381 from Beth Shean. Drawing by A. M. Pollastrini. (Not to scale)	83
4.16	Copper alloy armour scale from Beth Shean (Stratum Q-2/Level Late VII; Locus 889018). Drawing by A. M. Pollastrini. (Not to scale)	84
4.17	Copper alloy armour scale bearing the cartouche of Ramesses II from Kanakia, Salamis Island. Drawing by A. M. Pollastrini. (Not to scale)	85
4.18	Rawhide armour scale from Tell el-Amarna. Drawing by A. M. Pollastrini. (Not to scale)	87
5.1	Evolution of the Egyptian terms related to the different parts of the defensive panoply (A. M. Pollastrini)	89

Preface and Acknowledgements

The book you are holding in your hands is the outcome of many years of research and study. It derives largely from my doctoral dissertation, *L'armement défensif personnel égyptien pendant le Nouvel Empire* (École Pratique des Hautes Études, Paris) 2021, but not just from there. My researches on warfare, military history, heraldry and hoplology have their roots in a period of my life that predates the beginning of my 'doctoral journey'. These fields of study have fascinated me for as long as I can remember. However, they are not the most popular disciplines to deal with, and prejudice still surrounds them. The many unpleasant epithets given to me by short-sighted teachers due to my passion for these subjects are still in my mind.

This study is intended to be a guide to the Late Bronze Age coverings worn to protect the body in battle and the problems connected with their production and diffusion in Egypt. The book is aimed at a general audience interested in ancient warfare. The presumption is made that the reader will have a slight, but not extensive, knowledge of a glossary of armour.

As might be expected, such historical and technical work can only be achieved with the support and the help of many. Among those to whom I owe the greatest debt of gratitude for their support for this project over the years, I must thank, first and foremost, my advisor, Pascal Vernus, who always believed in me during my PhD period and subsequently set this book in motion. I am also grateful to the members of my dissertation committee, Marc Gabolde, Juan Carlos Moreno Garcia and Andreas Stauder, for their kind recommendations, which have been very useful during the writing process for this book.

I would like to express my sincere gratitude to Lily Mac Mahon, Zoe Osman and Georgina Leighton at Bloomsbury, who helped shape and produce the book you are holding in your hands.

I also owe a great debt to my friend and mentor, Paolo Gallo of the University of Turin, for the excellent training in Egyptology I received under his direction. His teachings are at the base of my current devotion to the ancient civilizations of the Nile.

To many others who kindly lent their expertise and technical support over the years, I would like to say 'thank you' to Stephen Harvey, Raymond Johnson, Anthony Spalinger and André Veldmeijer for providing me with helpful advice and often unpublished data.

Librarians and archivists have been central to the completion of this study. I am extremely grateful to Silvia Mosso of the Egyptian Museum of Turin for her willingness to help me venture into the labyrinth of bibliographic references.

I would also like to thank Guido Guiducci and Stefano Valentini of the Center for Ancient Mediterranean and Near Eastern Studies at Florence for allowing me to use the Center's library.

Last but not least, I would like to thank my family, especially my wife Ilaria and my daughters Morgana and Ginevra, for always being at my side, even in my darkest hour, and encouraging me to do my best. Without their love and support, this work would not have been possible.

Note for the Reader

All the Pharaonic regnal years mentioned in the text have been taken from E. Hornung, R. Krauss and D. A. Warburton, eds, *Ancient Egyptian Chronology* (*HbOr* 83), Leiden, Boston: Brill (2006).

The Western Asian chronology has been taken from M. Liverani, *Antico Oriente: storia, società, economia*, Bari: Laterza (1988).

For what concerns the transliteration of the Ancient Egyptian excerpts included in this book, I employed the Leiden Unified Transliteration established at the Thirteenth International Congress of Egyptologists (ICE) and accepted by the members of the International Association of Egyptologists (IAE) as the new standard in the transliteration/transcription of the Ancient Egyptian language.

For the abbreviations used in this work, see B. Mathieu, *Abréviations des périodiques et collections en usage à l'Institut français d'archéologie orientale*, Cairo: IFAO (2010).

Abbreviations

ABSA	*Annual of the British School of Athens.* Inst. of Class. Stud. (London).
AcArch (C)	*Acta archaeologica* (Copenhagen).
ADOI	*Assyrian Dictionary of the Oriental Institute.* University of Chicago (Chicago, IL).
AfrBull	*Africana Bulletin.* Univ. of Warsaw (Warsaw).
ÄgAbh	*Ägyptologische Abhandlungen* (Wiesbaden).
ÄgLev	*Ägypten und Levante / Egypt and the Levant. Zeitschrift für ägyptische Archäologie und deren Nachgebiete* (Vienna).
AJA	*American Journal of Archaeology.* Archaeol. Inst. of Amer. (New York, Baltimore).
AJIL	*The American Journal of International Law.* American Society of International Law (USA).
ANET	J. B. Pritchard, ed., *Ancient Near Eastern Text Relating to the Old Testament*, Princeton, New Jersey: Princeton University Press (1954).
AOAT	*Alter Orient und altes Testament* (Kevelaer, Neukirchen-Vluyn).
AOF	*Archiv für Orientforschung. Internat. Zeitschr. Für die Wiss. vom Vorderen Orient* (Berlin, Graz).
AOS	*American Oriental Series* (New Haven, Conn.).
APF	*Archiv für Papyrusforschung und verwandte Gebiete* (Leipzig, Stuttgart).
ArchAnz	*Archäologischer Anzeiger. Jahrb. des deutsch. archäol. Inst.* (Berlin).

ArchClass	*Archeologia classica. Riv. della Scuola naz. di archaeol.* (Rome).
ArchRep	*Archaeological Reports.* Council of the Society for the Promotion of Hellenic Studies and the British School at Athens (London).
ASAE	*Annales du Service des Antiquités de l'Égypte* (Cairo).
AuOr	*Aula Orientalis. Rev. de estud. del Proximo Oriente antiguo* (Barcelona).
BAAL	*Bulletin d'Archéologie et Architecture Libanaises* (Beirut).
BAR	*Biblical Archaeology Review.* Bibl. Archaeol. Soc. (Washington, DC).
BAR-IS	*British Archaeological Reports, Internat. Series* (London).
BASOR	*Bulletin of the American Schools of Oriental Research in Jerusalem and Baghdad* (Ann Arbor, MI, New Haven, CN).
BCH	*Bulletin de correspondence hellénique* (Paris).
BiAeg	*Bibliotheca aegyptiaca* (Brussels).
BIE	*Bulletin de l'Institut égyptien,* puis *Bulletin de l'Institut d'Égypte* (Cairo).
BiEg	*Bibliothèque égyptologique comprenant les œuvres des égyptologues français dispersées dans divers recueils et qui n'ont pas encore été réunies jusqu'à ce jour* (Cairo, Paris).
BiEtud	*Bibliothèque d'étude.* Inst. Franç. d'archéol. orient (Cairo).
BIFAO	*Bulletin de l'Institut français d'archéologie orientale* (Cairo).
BiOr	*Bibliotheca orientalis.* Nederlands Inst. voor het Nabije Oosten (Leiden).
BMFA	*Bulletin of the Museum of Fine Arts* (Boston, MA).
BMMA	*Bulletin of the Metropolitan Museum of Art.* Metropol. Mus. (New York).

BMMNEA	*Bulletin of the Museum of Mediterranean and Near Eastern Antiquities.* Medelhavsmuseet (Stockholm).
BSAK	*Studien zur altägyptischen Kultur – Beihefte* (Hamburg).
BSEG	*Bulletin de Société d'égyptologie de Genève* (Geneva).
BSFE	*Bulletin de la Société française d'égyptlogie* (Paris).
CahKarn	*Cahiers de Karnak.* Centre franco-égyptien d'étude des temples de Karnak (CFEETK), Centre nat. de la rech. sc. (Paris).
CDD	*Chicago Demotic Dictionary,* https://oi.uchicago.edu/research/publications/demotic-dictionary-oriental-institute-university-chicago
CEDAE	Centre d'étude et de documentation sur l'Ancienne Égypte (Cairo, Paris).
CGC	*Catalogue général du musée du Caire* (Cairo).
CRAIBL	*Comptes rendus de l'Académie des inscriptions et belles-lettres* (Paris).
CSEG	*Cahiers de la Société d'Égyptologie* (Geneva).
DemGl	W. Erichsen, *Demotische Glossar,* Copenhagen: Ejnar Munksgaard (1954).
DemStud	*Demotischen Studien* (Leipzig, then Würsberg).
EgArch	*Egyptian Archaeology.* Egypt Explor. Soc. (London).
EgMa	*Égypte et monde arabe.* CEDEJ – Centre d'études et de documentation économiques juridiques et sociales (Cairo).
Egypte	*Egypte. Afrique et Orient.* Centre vauclusien d'égyptologie (Avignon).
ENiM	*Égypte nilotique et méditerranéenne* (Montpellier).
ErIsr	*Eretz Israel. Archaeol. and Geogr. Stud.* Bialik Inst. (Jerusalem).

EVO	*Egitto e Vicino Oriente. Riv. della sez. orient. Ist. di stor. antica*, Universty of Pisa (Pisa).
FIFAO	*Fouilles de l'Institut français d'archéologie orientale* (Cairo).
GOF	*Göttinger Orientforschungen* (Wiesbaden).
GöttMisz	*Göttinger Miszellen. Beitr. zur ägyptol. Diskuss.* (Göttingen).
HÄB	*Hildesheimer ägyptologische Beiträge* (Hildesheim).
HbOr	*Handbuch der Orientalistik* (Leiden, Köln).
IEJ	*Israel Exploration Journal.* Israel Explor. Soc. (Jerusalem).
Iraq	*Iraq.* Brit. School of Archaeol. in Iraq (London).
JAEA	*The Journal of Ancient Egyptian Architecture* (open acess periodical).
JAOS	*Journal of the American Oriental Society* (New Haven, CN).
JARCE	*Journal of the American Research Center in Egypt* (Boston, New York).
JCunStud	*Journal of Cuneiform Studies* (Ann Arbor, Mich., New Haven, CN).
JEA	*Journal of Egyptian Archaeology.* Egypt Explor. Soc. (London).
JEOL	*Journal jaarbericht van het Vooraziatisch – Egyptische Genootschap (Gezelschap) 'Ex Oriente Lux'* (Leiden).
JGS	*Journal of Glass Studies.* Corning Museum of Glass (New York).
JKAF	*Jahrbuch für kleinasiatische Forschung* (Heidelberg).
JNES	*Journal of Near Eastern Studies.* Dept. of Near Eastern Lang. and Civilis., Univ. of Chicago (Chicago, IL).

JournAs	*Journal asiatique.* Soc. asiat. de Paris (Paris, then Leuven).
JRGZ	*Jahrbuch des römisch – germanischen Zentralmuseums* (Bonn).
JSSEA	*Journal of the Society of the Studies of Egyptian Antiquities* (Toronto).
Kêmi	*Kêmi. Rev. de phil. et d'archéol. égypt. et copte* (Paris).
KMT	*K.M.T. A Modern Journ. of Anc. Egypt* (San Francisco, CA).
KRI	K. A. Kitchen, *Ramesside Inscriptions*, 1969–90 (Oxford).
LÄ	*Lexikon der Ägyptologie*, 7 vols, ed. W. Helck, E. Otto, W. Westendorf, 1972–92 (Wiesbaden).
LD	K. R. Lepsius, *Denkmäler aus Ägypten und Äthiopien*, 1849–59 (Berlin).
MÄS	*Münchner ägyptologische Studien* (Berlin, Munich).
MDAIK	*Mitteilungen des deutschen Instituts, Abt. Kairo* (Wiesbaden).
Memnonia	*Memnonia.* Assoc. pour la sauvegarde du Ramesseum (Paris).
Mesopotamia	*Mesopotamia. Rivista di Archeologia, Epigrafia e Storia Orientale Antica.* University of Turin (Turin).
MIFAO	*Mémoires publiés par les membres de l'Institut français d'archéologie orientale* (Cairo).
MMAEE	*Metropolitan Museum of Art, Egyptian Expedition. Metropol. Mus.* (New York).
MMAF	*Mémoires publiées par les membres de la Mission archéologique française au Caire.* Inst. franç. d'archéol. Orient (Cairo).
MPER	*Mitteilungen aus der Papyrussammlung der österreichischen Nationalbibliothek (Papyrus Erzherzog Rainer).*

NARCE	*Newsletter of the American Research Center in Egypt* (Princeton, Cairo).
OBO	*Orbis biblicus orientalis* (Fribourg, Switzerland, Göttingen).
OIP	Oriental Institute Publications, University of Chicago (Chicago, IL).
OIR	Oriental Institute, Annual Reports. University of Chicago (Chicago, IL).
OJA	*Oxford Journal of Archaeology.* Oxford University (Oxford).
OLA	*Orientalia Analecta Lovaniensia.* Dept. Orient. (Leuven).
Orientalia	*Orientalia. Comment. periodici Pontific. Inst. biblici* (Rome).
Paléorient	*Paléorient. Rev. pluridisc. de préhist. et de protohist. de l'Asie du Sud-ouest.* Centre nat. de la rech. scient. (Paris).
ParPass	*La Parola del passato. Riv. di stud. antichi* (Naples).
Picus	*Picus. Stud. e ric. sulle marche nell'Antich.* (Macerata).
PM	Porter (B.), Moss (R. L. B.), *Topographical Bibliography of Ancient Egyptian Hieroglyphic Text, Reliefs and Paintings* (Oxford).
PN	H. Ranke, *Die ägyptischen Personennamen*, 3 vols, Glückstadt / Hamburg: Verlag J. J. Augustin (1935–77).
ProblÄg	*Probleme der Ägyptologie* (Leiden).
RAAO	*Revue d'assyriologie et d'archéologie orientale* (Paris).
RecTrav	*Recueil de travaux relatifs à la philologie et à l'archéologie égyptiennes et assyriennes* (Paris).
ResAnt	*Res Antiques* (Bruxelles).
RLA	*Reallexicon der Assyriologie* (Berlin, Leipzig).
SAAB	*State Archives of Assyria Bulletin* (Padua).
SAK	*Studien zur altägyptischen Kultur* (Hamburg).

SAOC	*Studies in Ancient Oriental Civilizations* (Chicago, IL).
Scienze dell'Antichità	*Scienze dell'Antichità. Storia, Archeologia, Antropologia.* Dep. of Science of Antiquities, Sapienza Univ. (Rome).
ShirEgypt	*Shire Egyptology* (Aylesbury).
Studia Eblaitica	*Studia Eblaitica. Studies on the Archaeology, History, and Philology of Ancient Syria.* Sapienza Univ. (Rome).
Sumer	*Sumer. Journ. of Archaeol. and Hist. in Arab Word* (Baghdad).
Syria	*Syria. Rev. d'art orient. et d'archéol.* (Paris).
Talanta	*Talanta. Proceedings of the Dutch Archaeol. and Hist. Soc.* (Groningen).
Tel Aviv	*Tel Aviv. Journ. of the Tel Aviv Univ. Inst. of Archaeol.* (Tel Aviv).
TrabEg	*Trabajos de Egiptología* (Madrid).
Urk	*Urkunden des ägyptischen Altertums* (Leipzig, Berlin).
VicOr	*Vicino Oriente. Annuario del Dipart. di sc. stor. archeol. e antropol. dell'Ant.* Univ. degli stud. (Rome).
VIO	*Veröffentlichungen der deutschen Akademie der Wissenschaften zu Berlin des Institut für Orientforschung* (Berlin).
Wb	Erman (A.), Grapow (H.), *Wörterbuch des aegyptischen SpracAhen* (Leipzig, Berlin).
WZKM	*Wiener Zeitschrift für die Kunde des Morgenlandes.* Verb. der wissenschsftl. Gesellsch. Österreichs (Vienna).
ZÄS	*Zeitschrift für ägyptische Sprache und Altertumskunde* (Leipzig, Berlin).
ZeitAss	*Zeitschrift für Assyriologie* (Heidelberg).
ZDPV	*Zeitschrift des deutschen Palästina – Vereins.* Deutsch. evang. Inst. für Altertumswiss des Heiligen Landes (Wiesbaden).

1

Introduction

1.1 The scope, methodology and purpose of the present study

The purpose of this book is to examine the dynamics related to the introduction and diffusion of protective clothing in Egypt during the New Kingdom (Eighteenth to Twentieth Dynasties, sixteenth to eleventh centuries BCE). The word 'introduction' represents perhaps the best term to define this phenomenon because this type of military equipment is not an Egyptian technological innovation. Contrariwise, armours appeared at the end of the Bronze Age, following the gradual Hurrian–Mitanian expansion in Western Asia, and then spread throughout the surrounding territories, including Egypt. The development, during three centuries of Mitanian rule, of a new way to fight, based on the use of masses of horse-drawn war chariots, indeed encouraged the adoption of a type of personal defence equipment that was intended to provide the advantage of leaving hands free to wield weapons and shields, and drive vehicles. So, we can assert that for members of chariot crews the armour represented the best answer to the need for personal protection on the battlefield.

John Keegan has lucidly described the terrible impact of the new war chariot tactics on the Western Asiatic armies, mainly composed of unarmoured infantrymen:

> Circling at a distance of 100 or 200 yards from the herds of unarmoured foot soldiers, a chariot crew – one to drive, one to shoot – might have transfixed six men a minute. Ten minutes' work by ten chariots would cause 500 casualties or more, a Battle of the Somme-like toll among the small armies of the period. In the face of such an attack by an enemy against which it could

not manoeuvre out of trouble, the stricken host had only two choices: to break and run or to surrender.[1]

Unlike other foreign coeval weapons and equipment – such as the *khopesh* sickle-shaped sword, the composite bow, and the chariot – helmets and body armour have not been adopted as attributes in the pharaonic ideology or within the context of religious beliefs in any meaningful way. Only the Ramesside texts related to the battle of Kadesh and the siege of the Syrian town of Dapur mention corslets in strict relation to the fighting Pharaoh. The references to protective clothing associated with a specific deity or worship are currently inadequate to achieve convincing results.

I have paid particular attention to the survey of pieces of evidence and the creation of a corpus broadly organized into three sections dedicated to the iconographic, archaeological and lexicographic attestations, respectively. Based on the collated information, I have tried to develop the most accurate perspective on how the helmet and the cuirass were introduced and propagated in Egypt, not forgetting their impact on warfare and their possible role in ideology. Moreover, in this work, I have adopted an approach involving not only recurrent recourse to comparison between Egyptian pieces of evidence and the contemporary attestations coming from the Middle East and the Aegean region but also the introduction in the text of some excursus and digressions about the defensive equipment related to ages that fall outside the restricted chronological limits of New Kingdom.

Of course, some important questions remain unanswered and readers should, therefore, not be surprised by the several open questions that they will find in this text. We might give a better explanation of that by borrowing the words of Doyne Dawson:

> Archaeology can tell us much about arms and armour, for some periods more than others, but never about the use of the weapons, which is where the real problem lies.[2]

Finally, readers should be aware that they are not encountering a practical handbook for armour makers or historical reenactors. Technical aspects of the manufacture of Late Bronze Age protective clothing have been largely ignored, except in the parts of the text where they are strictly necessary. Perhaps some readers may be disappointed by this revelation, but I felt it was better not to repeat what has been covered comprehensively elsewhere. However, to find a way to make amends for my fault, I dedicated a part of the next section, '1.2 Previous studies', to several academic publications, which, in whole or in part, deal with armour structure and the various stages of construction.

1.2 Previous studies

In the context of the relatively marginalized ancient Egyptian military and warfare studies, helmets and armour have surprisingly caught the attention of scholars since the dawn of modern Egyptology at the end of the eighteenth century. However, a thorough survey of the sources of evidence for the origin and spread of Egyptian body protections during the New Kingdom has never been carried out until now.

Commissioned by General Napoleon Bonaparte, the *Description de l'Égypte* (full title: *Description de l'Égypte, ou Recueil des observations et des recherches qui ont été faites en Égypte pendant l'expédition de l'armée française*) is rightly deemed to be the first scientific survey of Egypt. This work, collected in twenty-three volumes, is the outcome of the collaborative research of several dozen scholars who accompanied the French Campagne d'Égypte from 1798 to 1801, including contributions that focus on Egyptian antiquities, natural history and contemporary customs and traditions. In the volume *Antiquités, Planches. Tome II* (1812) for the first time a full-colour plate (Figure 1.1)[3] shows a helmet and a scale cuirass copied from the walls of Room M in the Theban tomb of Ramesses III (KV 11). A few years later, all offensive and defensive weapons depicted in the tomb of Ramesses III have been fully reproduced in Jean-François Champollion's *Monuments de l'Égypte et de la Nubie. Planches. Tome III*, pl. CCLXII–CCLXIV (1845).

The collection of plates entitled *Monuments égyptiens, bas-reliefs, peintures, inscriptions, etc., d'après les dessins exécutés sur les lieux* (1847) follows in the wake of the former work. In the preface of his book, Émile Prisse d'Avennes indeed stated that he published his drawings with the purpose of 'combler une partie des lacunes qui se trouvent dans le recueil de Champolion le jeune: *Monuments de l'Égypte et de la Nubie*'. The details of New Kingdom reliefs shown in plate 35 and even in colour plate 46 could prove that scholars became more interested in ancient Egyptian military equipment, as Egyptological

Figure 1.1 Military standards, weapons and other objects from the tomb of Ramesses III (KV 11). *Description de l'Égypte: Antiquités, Planches.* Tome II, pl. 88.

studies advanced. However, if on the one hand, the above-mentioned works of Champollion and Prisse d'Avennes can be perceived as signs of an early methodological approach to Egyptian hoplology, on the other hand, the two authors only provided brief descriptions of the monuments and objects examined, just emphasizing their artistic side.

Even though John Gardner Wilkinson's pioneering work *The Manner and Customs of the Ancient Egyptians* (1837, 1841) is not entirely focused on ancient Egyptian warfare, it features, nonetheless, an exhaustive digression about military equipment, which remains a significant milestone in the history of Egyptian warfare studies. The 1878 edition of *Manner and Customs,* revised and augmented by Samuel Birch, contains, in particular, a detailed description of the armour components, widely based on iconographic sources. Moreover, an extended corpus of artworks drawn by the author himself accompany the written text. Nevertheless, on a closer examination, the material presented here lacks any reference to Egyptian chronology. And in the last instance, Wilkinson made several inspiring comparisons between Pharaonic and ancient Greek weapons, in order to find parallels between them. Unfortunately, the latest arms and armour studies have not confirmed most of the observations he made in his work.

Since its publication in 1926, Walther Wolf's *Die Bewaffnung des altägyptischen Heeres* has been deservedly considered the first remarkable reference manual on ancient Egyptian military and warfare. This text is an admirable account of the knowledge on the subject gathered up to that time. Due to its structure built on a reliable chronology and its remarkable iconographic corpus, it is essential for approaching the study of Pharaonic hoplology even today. This publication includes a short section devoted to the armour components and the related ancient Egyptian terminology.

In the very same year, Hans Bonnet published *Die Waffen der Völker des alten Orients* (1926), in which he tried to summarize the relations between Egypt and ancient Western Asian cultures, through the shape and the use of weapons, including helmets and body armour. Bonnet did not forget to make informative comparisons with the weaponry peculiar to some modern peoples. Nevertheless, as the author himself stated, the volume lacks a proper conclusion due to the scarcity of documentary source material and its mostly fragmentary nature.

Yagael Yadin's *The Art of Warfare in Biblical Lands* (1963) follows the steps of the former book. This two-volume study surveys military equipment and tactics in Anatolia, Syria, Mesopotamia, Palestine and Egypt from prehistory to the destruction of the First Jerusalem Temple by the Neo-Babylonian Empire (587 BCE). The author, a renowned biblical archaeologist and former officer in the Israel Defence Forces combined his specialized competencies in archaeology and military matters in order to provide the reader with a chronologically arranged account of the principal weapons, armour, fortification and tactics. Two entire pages, filled with photographs of metallic armour scales and drawings of coats of scales from the Egyptian reliefs, are displayed in the large body of pictures and artworks, which represents one of the strengths of this monograph. At the same time, however, the lack of notes or indexes to the text makes this work more suitable for the general public and less reliable for scholars.

Starting from the second half of the last century, the increasing interest in Late Bronze Age military technology has encouraged the publication of several studies focused on the spreading of personal protective equipment in the Eastern Mediterranean basin.

The entries 'Helm' by Rolf Krauß in *LÄ*, Band 2 (1977), 1114–15, and 'Panzer(hemd)' by Wolfgang Decker in *LÄ*, Band 4 (1981), 665–6, represent an attempt to summarize the iconographic, archaeological and lexicographic pieces of evidence concerning Egyptian helmets and armour. Because of the structure, these references are naturally brief. Although nowadays the two entries look outdated, they still are a good starting point for anyone who wishes to deepen their study of ancient Egyptian hoplology.

For what concerns the helmets, Timothy Kendall's 'gurpisu ša awēli: The Helmets of the Warriors of Nuzi' provides an in-depth analysis of the Eastern Mediterranean head protections from the second half of the fifteenth to the first half of the fourteenth century BCE. In this essay, which is part of the edited volume *Studies on the Civilization and Culture of Nuzi and the Hurrians, Vol. 1* (1981), particular attention has been paid to Nuzi inventories mentioning suits of armour or part of them. Ephraim Avigdor Speiser had previously examined Nuzi armour terminology in the brief communication entitled 'On Some Articles of Armor and Their Names', *JAOS*, Vol. 70, No. 1 (1950), 47–9. However, Kendall does not refer specifically to the almost coeval depictions of Asiatic helmets in the Amarna art. Alan Schulman has instead addressed this aspect in the article called 'Some Observations on the Military Background of the Amarna Period', *JARCE* 3 (1964), 51–69 and in 'Hittites, Helmets and

Amarna: Akhenaten's First Hittite War', which is part of the collective work *The Akhenaten Temple Project, Vol. 2: Rwd – mnw, Foreigners and Inscriptions* (1988).

Tamás Dezső has long been interested in the history of the Assyrian army. His most prominent work, *Near Eastern Helmets of the Iron Age* (2001), is an in-depth survey of Western Asiatic types of helmet based on textual, iconographic and archaeological sources, dating back to the first millennium BCE. Moreover, protective gear consisting of scales is the focus of his study called 'Scale Armour of the Second Millennium BC' (2002). This article is an invaluable reference, which gathers archaeological, iconographic and written data from Mesopotamia, Iran, Urartu, Anatolia, Syria, Phoenicia, Palestine, Egypt and Cyprus. Moreover, it provides the reader with two useful tables where all the data concerning the scale helmets and armour from the Nuzi inventory lists are recorded. A few years later, Dezső returned to this topic with the entry 'Panzer' in Band 10 (2004) of *Reallexicon der Assyriologie und Vorderasiatischen Archäologie*. This comprehensive survey of body armour dating back to a period ranging from the second to early first millennium BCE is nevertheless to be completed with the latest finds from Greece and Western Asia.

Special mention should be given to Fabrice De Backer's study 'Evolution of the Scale Armour in the Ancient Near East, Aegean and Egypt: An Overview from the Origins to the Pre-Sargonids', *ResAnt* 8 (2011). De Backer starts his article with a systematic review of the different types of armour design, aiming to provide the reader with the tools to identify differences and similarities of the protection represented or mentioned in the Western Asian, Aegean and Egyptian pieces of evidence. Then, the author gives a chronological and geographical classification of various archaeological, iconographic and written sources available to him. In addition to the article mentioned above, De Backer explores in detail the use of scale armour during the first millennium BCE – especially in the context of the Neo-Assyrian Empire – in some further publications: 'Siege-Shield and Scale Armour Reciprocal Predominance and Common Evolution', *Historiae* 8 (2011); 'Scale-Armours in the Neo-Assyrian Period: A Survey', *SAAB* XIX (2011–12); *Scale-Armours in the Neo-Assyrian Period: Manufacture and Maintenace* (2013) and 'Une armure expérimentale du Premier Millenaire av. J.–C.' (2015).

As regards the technical aspects of the Bronze Age scale armour manufacture, we cannot omit mention of Bengt Thordeman's pivotal study: 'The Asiatic Splint Armour in Europe', *AcArch* (C) 4 (1993). Following the excavation of the mass grave from the Battle of Visby on the island of Gotland (Sweden), various fourteenth-century armour splints have been brought to light. These fragments were considered the latest evidence of lamellar armour in Europe. Thordeman consequently listed a range of records of scale and lamellar protections from Antiquity to the Middle Ages, considering related lacing techniques from Antiquity to the Middle Ages to find possible parallels with the armour splints from Visby. Later, the Swedish archaeologist returned to this topic, devoting a whole chapter of his monograph entitled *Armour from the Battle of Wisby 1361* (1939) to the history of the lamellar protections in the Eurasian continent, extending the survey to a larger quantity of scales and splints from the Eurasian continent.

Walter Ventzke has attempted to reconstruct virtually Late Bronze Age body armour, following the discovery of a set of 180 bronze scales and 40 bronze staples excavated in Kāmid el-Loz, in the Lebanese Beqaa plain. In 'Zur Rekonstruktion eines bronzenen Schuppenpanzer', which is a part of the exhibition catalogue *Frühe Phöniker im Libanon: 20 Jahre deutsche Ausgrabungen in Kāmid el-Loz* (1983), Ventzke postulated the method by which the scales of various shapes should have been attached to each other.

The doctoral dissertation *Late Bronze Age Scale Armour in the Near East: An Experimental Investigation of Materials, Construction, and Effectiveness, with Consideration of Socio-economic Implications* (2002) by Thomas Hulit stands out for the attention devoted to the collection of a large corpus of archaeological data concerning the assembling of Late Bronze Age scale armour. The virtual reconstruction of Tutankhamun's body armour, suggested by Hulit after an extensive autoptic examination of what remains of the corslet, is undoubtedly the most outstanding part of this work.

More recently, the technical aspects of the manufacture of Late Bronze Age coats of armour have been the subject of two other publications: *Bronze Age Military Equipment* (2011) by Dan Howard and *Armour Never Wearies: Scale and Lamellar Armour in the West, from the Bronze Age to the 19th Century* (2013) by Timothy Dawson.

As regards Tutankhamun's body armour, André Veldmeijer, Thomas Hulit, Lucy-Anne Skinner, and Salima Ikram have recently published the article 'Tutankhamun's Cuirass Reconsidered', *JEOL* 48 (2021–2), which is so far the most comprehensive study on the subject. In that work, the above-mentioned observations on Tutankhamun's body armour, made by Thomas

Hulit in his doctoral dissertation, have been updated in the light of the results obtained from the latest archaeometry techniques.

Among the several works aimed at a popular audience, we must mention at least two books produced by Osprey Publishing, one of the leading publishing companies in illustrated military history. *Ancient Armies of the Middle East* (1981) by Terence Wise and *New Kingdom Egypt* (1992) by Mark Healy focus on the organization. They both contain full-colour artwork plates by Angus McBride. Even though the renowned illustrator of military subjects succeeded in giving a vivid impression of the clothing and equipment of New Kingdom Egyptian soldiers and their enemies, he sometimes adopted a speculative approach to creating his historical illustrations, affecting the credibility of the outcome.

Lastly, although not strictly pertinent to the subject of this study, we cannot avoid referring to Marianne Mödlinger's contribution to scholarship in the fields of hoplology and archaeometallurgy. Among Mödlinger's many publications concerning the manufacture and usage of metal body armour in Bronze Age Europe, we must mention at least two articles of particular interest for the present study: 'European Bronze Age Cuirasses: Aspects of Chronology, Typology, Manufacture and Usage', *JRGZ* 59 (2012) and 'From Greek Boar's-Tusk to the First European Metal Helmets: New Approaches on Development and Chronology', *OJA* 32(4) (2013). A large part of these studies were then subsumed into a monograph entitled *Protecting the Body in War and Combat: Metal Body Armour in Bronze Age Europe* (2017).

1.3 Early evidence of helmets and body armour in Western Asia and Egypt

When did early humans begin to use protective clothing or equipment designed to protect the wearer's body from combat injuries? To provide an answer to this question, we might hazard a guess that the use of defensive covering for the body extends back beyond historical records when prehistoric warriors protected themselves with clothing made of natural raw materials.

To better understand the dynamics of the diffusion of helmets and body armour in Egypt during the New Kingdom, it is first necessary to broaden the chronological and geographical horizons of the research. As offensive weapons evolved to become more effective in inflicting physical injuries upon the enemy, so did the helmets and armour used to deflect or absorb the impact of those weapons. That is because the evolution of protective equipment should be considered as the result of a long and articulated development.

1.3.1 Western Asia

The earliest forms of protective equipment are attested in Mesopotamia at least from the third millennium BCE. Sir Leonard Woolley discovered the earliest currently known specimens of copper alloy helmet at Ur (located in modern-day Iraq near the city of Nasiriyah) in the tomb PG 789, known as the 'King's Grave'.[4] The structure was part of a group of sixteen graves, called 'Royal Tombs', which are dated to the early ED IIIa (*c.* 2600–2450 BCE). The remains of six sacrificed bodyguards wearing helmets and carrying spears with copper heads were found in the *dromos* of the tomb (Figure 1.2). The six helmets are

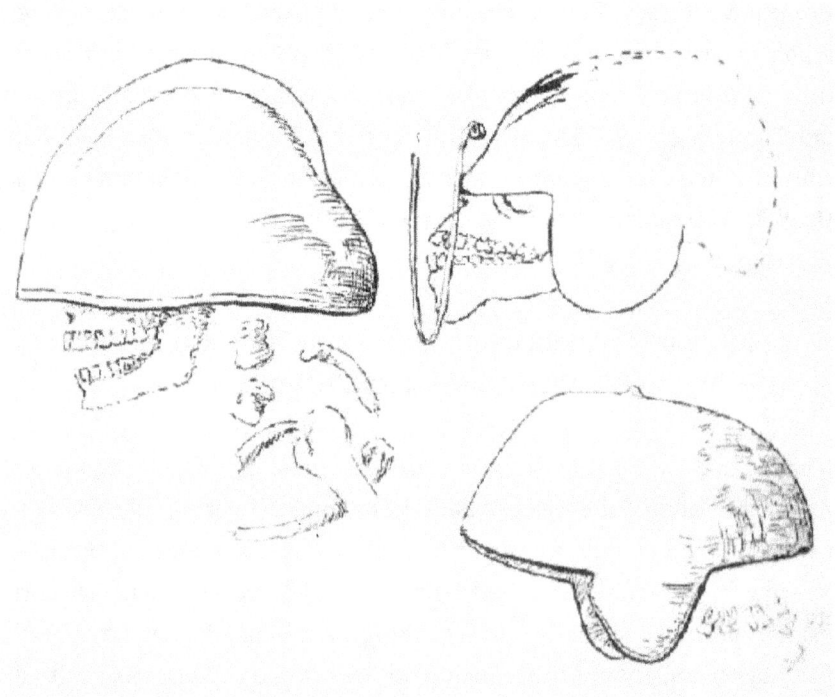

Figure 1.2 *In-situ* drawing of helmets on crushed skulls from PG 789 tomb in the Royal Cemetery at Ur. Woolley, C. L., *The Royal Cemetery*, pl. 218 *a-b-c*.

currently preserved in the British Museum. Some scholars have identified two distinct types of helmets, entirely produced in hammered sheet metal. Among the six helmets found in the tomb PG789, a globular skull with an uppermost button and cheek guards characterizes four of them. A round skull decorated with an axial ridge distinguishes the two remaining helmets from the others. Recent radiographic examinations on the remains of at least three soldiers from tomb PG 789[5] have revealed a series of small holes following the edges of the cheek guards and the lower edge of the helmets. These holes were probably used to fasten a lining, inside and out, and overlapping at the outer edge.

At the beginning of the twentieth century, scattered pieces of a contemporary copper alloy helmet were found during excavations by the French archaeologist Gaston Cros at the archaeological site of Tello (the ancient city Girsu) in what is today southern Iraq. The helmet, apparently dating back to the Early Dynastic III, did not look much different from the above-mentioned Ur headdresses.[6] According to the reconstruction drawing made by Cros in 1906 (Figure 1.3), the slightly conical skull of this headdress extends down the nape of the neck and covers the ears. A series of small holes have been made along the lower forward edge of the helmet, perhaps to attach a chinstrap to it. What remains of the Tello helmet is currently kept in the Louvre Museum in Paris (AO, 4119).

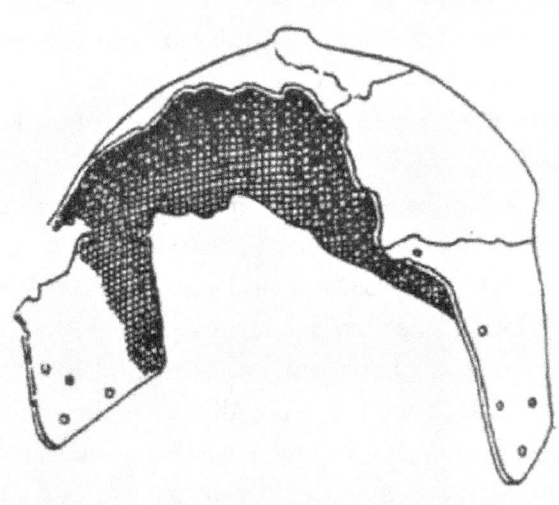

Figure 1.3 Reconstruction drawing of the copper alloy helmet from Tello. Cros, G., *RAAO*, Vol. 6, No. 3 (1906), 89.

There is no lack of iconographical sources about the Mesopotamian employment of body protections during the third millennium BCE. Perhaps the finest of these is the so-called 'Standard of Ur': an originally wooden box decorated on four sides with a mosaic made from shell, lapis lazuli and red limestone. This artefact was found by Woolley in tomb PG 779 in the Royal Cemetery at Ur and is currently on display at the British Museum (BM 121201).[7] One of the two rectangular side panels depicts a battle scene arranged in three registers. The Sumerian foot soldiers and charioteers wear tight-fitting, open-faced headdresses of slightly conical form, fastened by a strap under the chin. The lower edge of these helmets extends down the nape of the neck to protect the back of the head, the rear of the neck and the ears. Is it reasonable to assume that the crushed helmets found in tomb PG 789 at Ur had approximately the same appearance? This question is not easy to answer in precise detail because the helmets depicted on the 'Standard of Ur' are generally considered to be made of leather.[8]

A long spotted cloak fastened at the neck completes the soldiers' defensive equipment. The true nature of this loose garment, known as 'bishop mantle' in academic literature, has not yet been fully clarified. Y. Yadin first and then F. De Backer interpreted it as a leather cloak reinforced with circular or hemispherical metallic studs.[9] D. Howard, in contrast, rejected this speculation, claiming that no metallic discs or studs were contained in the tombs of Ur. Moreover, some metallic elements would have increased the weight and the cost of the garment without substantially improving its effectiveness. He, in turn, states that the cloaks appear to have been made with the hide of an animal with dotted fur.[10]

The so-called 'Stele of the Vultures' is a unique visual representation of third millennium BCE Mesopotamian warriors in combat. This monument, dating back to the ED IIIb (*c.* 2450–2350 BCE), had been raised in the precinct of the temple of the god Ningirsu at Girsu to commemorate the victory of Eannatum, the third king of Lagash's third Dynasty (modern Tell al-Hiba, Iraq) over the city of Umma (modern Tell Jokha, Iraq). All that remains from the original stele is a group of seven fragments kept in the Louvre Museum (AO 50a-c + 2436-8 + 16109).[11] Considered by many scholars as one of the earliest surviving works of narrative relief sculpture, the white limestone slab shows a mythological scene on the obverse and a historical one on the reverse. The latter is divided into four registers representing the different stages of the victory of Eannatum over his enemies. A phalanx-like formation composed of eleven heavy infantrymen (nine men on the reverse face and two on the side) advancing on the naked corpses of the enemies is carved in the upper part of the stele. The soldiers in full fighting position are equipped with spears, battle axes, large rectangular shields covering the whole body, and helmets.[12] In the second register, a group of twelve soldiers is depicted just below the close-order formation mentioned above. Nevertheless, in this instance, the soldiers advance in loose order behind the royal war chariot. Except for the tall rectangular shields, the men are equipped in the same way as the previous ones: each soldier holds a long spear in his right hand and a battle axe in his left and wears a helmet. Upon closer examination, the headdresses engraved in the first and second register bear a strong resemblance to the copper alloy helmet found in the 'Maison des Fruits' at Tello (Figure 1.3). The helmets seem to be composed of a one-piece conical skull with a slightly pointed top. Their lower edge extends down the nape of the neck to protect the back of the head, the rear of the neck and the ears, leaving the face uncovered. Moreover, a further detail stands out. A marked line follows the edges of each headdress, perhaps representing the internal linen sewn all around the helmet edges or a padded coif worn under the helmet for comfort.

A relevant corpus of iconographic pieces of evidence related to the use of protective clothing during the second half of the third millennium BCE comes from pre-Sargonic Syria. The excavations undertaken at the sites of Mari (modern Tell Hariri, Syria) and Ebla (modern Tell Mardikh, Syria) have brought to light a large number of decorative elements and loose inlays relating to monuments made to celebrate war events with a technique very similar to that employed for the inlaid panels of Kiš (ED II, c. 2650–2550 BCE)[13] and the 'Standard of Ur'. After a quick overview of the available documentation – the fragments of the so-called 'Standard' from the temple of Ishtar of Mari,[14] the shell inlays from the Palace of Mari[15] and the marble inlays of the so-called 'Standard' from the Royal Palace of Ebla[16] – it can be assumed that a shared military system connected Mesopotamia and eastern Syria.[17]

Among the several fragments portraying armed warriors from Mari, a small decorative slab of limestone, found in 1968 by André Parrot in Room XLVI of the Palace P 1, appears to be the sole iconographic evidence of Early Dynastic advanced siege techniques.[18] The rectangular stone, currently kept in the Deir ez-Zor Museum (n°. 3746),[19] was probably part of a larger war scene. A spearman with a massive arched shield and an archer, aiming his (composite?) bow in an upward direction, figure prominently on the plaque.[20] Moreover, on the top left of the stone, the naked corpse of an enemy is depicted falling from the ramparts of the besieged city. The spearman and the bowman wear knee-length skirts and helmets similar to those shown on the 'Standard of Ur', but without with a chinstrap. The bowman also wears a sort of rectangular spotted cloak hanging loosely from his right shoulder, lending protection to the chest and back. Timothy Kendall suggests that the Nuzi word *tutiwa*, arguably derived from the Akkadian term *tutittu*, 'pectoral' (*ADOI*, Vol. 3, D, 168),[21] is probably related to this peculiar garment. Although its true nature is still unclear, the long rectangular cloak is quite frequently seen in contemporary representations of warriors from Mari and Ebla, together with the tight-fitting, open-faced Sumerian helmet.

The Akkadian official art provides several other unmistakable signs of the widespread use of helmets and armour.[22] The scenes of war carved on the fragmentary Victory stele of King Rimush (2278–2270 BCE) from Tello[23] and the well-known limestone stele of King Naram-Sim (2254–2218 BCE) from Susa (modern Shush, Iran)[24] show almost all the Akkadian warriors wearing helmets, which resemble those used during the Early Dynastic period. The headdresses are indeed characterized by a conical skull with a slightly pointed top. The lower edge of the skull is cut away at the front to form an arched face opening. Moreover, some of the warriors depicted on Rimush's Victory stele wear long rectangular scarves resembling the *tutiwa/tutittu* carved on the above-mentioned slab of limestone from Mari.

A completely different type of helmet is depicted on two fragments belonging to another Naram-Sim celebrative monument, currently kept in the Museum of Fine Arts of Boston and the Iraq Museum of Baghdad, respectively. The Boston fragment, MFA 66.893, and the Baghdad fragment, IM 55639, were part of an alabaster stele probably erected in the Nasiriyah region to commemorate one of Naram-Sin's victories in Anatolia.[25] A reconstruction of the front of the stele was attempted including a third fragment kept in the Iraq Museum (IM 59205), in addition to the two above-mentioned fragments.

Although extremely incomplete, the outcome shows that the decoration of the stele was arranged in superposed registers in which lines of bearded Akkadian warriors and naked prisoners of war are depicted. The equipment of the warriors seems relatively homogeneous. They wear long fringed skirts, broad sashes resembling those worn by the Akkadian warriors on the Rimush's Victory stele from Tello and helmets with horizontal grooves. The skulls of the helmets are round with a slightly pointed top. Each side of the skull is provided with a small ear guard. The rear edge extends down the nape of the neck to protect the back of the head.

The same type of helmet with horizontal stripes is carved on a small fragment of a stele excavated at Susa and currently kept in the Louvre Museum (Sb 6641 bis).[26] However, it is not easy to interpret the striated surface. Could it suggest that these helmets were produced with a quilting technique or, more simply, decorated with a striped pattern? The absence of archaeological evidence does not allow us to prove the use of a form of scale assembly and prevents us from formulating alternative assumptions.

1.3.2 Egypt

Compared with Mesopotamia and Syria, Egypt has provided a corpus of early attestations, which, after a quick overview, is to be considered more limited and controversial. From a chronological point of view, all the Egyptian pieces of evidence date back to the period from the end of the third millennium BCE to the beginning of the second millennium BCE.

The earliest potential evidence of the use of protective equipment in Egypt is given by a group of four copper plaques (Figure 1.4) found in the tomb of Mesehti 𓅓𓋴𓉔𓏏𓏭 at Asyut (PM IV, 256). Mesehti, a high official who lived between the late Eleventh and the early Twelfth Dynasty, is best known for the wooden models of spearmen and archers left in his tomb.[27] The four copper plaques, preserved at Petrie Museum in University College, London (n. UC38049A), measure approximately 8 x 3 centimetres. They have a rectangular shape with markedly rounded upper and lower edges. There are two lacing(?) holes roughly vertically aligned in each plaque. Even though the size and the overall appearance of the plaques are comparable to those of New Kingdom armour scales, the holes in the plaques are too few in number to

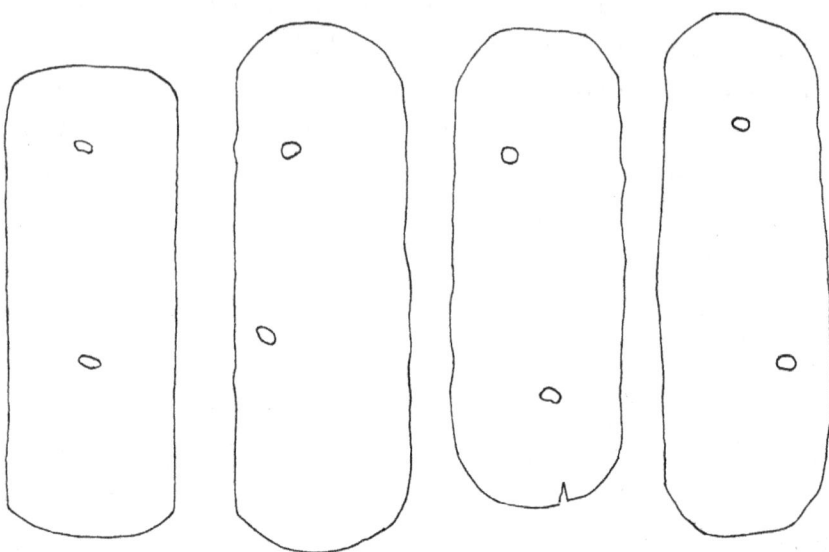

Figure 1.4 Copper plates found in the tomb of Mesehti, Asyut. Drawing by A. M. Pollastrini. (Not to scale.)

allow them to be laced both to each other and to a backing, as rightly suggested by Dan Howard.[28] This in itself does not reduce the uncertainty surrounding the actual function of the copper plaques. On the contrary, it makes their connection with protective equipment unlikely. However, if this connection were by any chance to be confirmed, the assumption that scale construction originated in Western or Central Asia would have to be reconsidered.

The painted decoration of the tomb of Djehutyhotep at Deir el-Bersha in Middle Egypt (PM IV, 179–181) reveals possible evidence of the early use of breastplates protecting the centre of warriors' chests.

Djehutyhotep (PN I, 408, 18) lived under the reigns of Amenemhat II (c. 1878–1843 BCE), Senusret II (c. 1845–1837 BCE), and Senusret III (c. 1837–1819 BCE) and was 'Great Chief (=nomarch) of the Hare Nome' (fifteenth *nomos* of Upper Egypt, also known as 'Ermopolite nome').[29] The tomb of Djehutyhotep is famous for the scene, unique in all Egyptian art, showing the transportation of the colossus statue of the deceased.[30] On the right wall of the inner chamber of the tomb, the owner, depicted on a larger scale than the other human figures around him, is observing with the members of his family the work of the farmers and artisans of his domain.[31] In the lower register of the decoration, the men of the entourage of Djehutyhotep are shown in a single line carrying weapons (spears, axes, bow and arrows and a massive shield) and other belongings (sandals, batons, a sedan chair and a casket) to their master. Among the retainers of the nomarch, two men wearing long skirts sport on their chests an oval element inserted within two crisscrossed straps. The same peculiar equipment is also worn by two figures depicted standing over Djahutyhotep in the above-mentioned scene of the transportation of a colossal statue. Again, they seem to be members of the deceased's retinue.[32] According to Bridget McDermott, these oval elements could be leather or quilted forms of armour used to protect the chest.[33] In light of this conjecture, can we assume that they are early examples of oval-shaped *kardiophylax*?[34]

Finally, we cannot omit mention of the discovery in 1923 of the unmummified remains of at least sixty men in rock tomb MMA 507 at Deir el-Bahari, on the West bank of Luxor (PM I, 650–651).[35] In 1923, when the team of the Egyptian Expedition of the Metropolitan Museum of Art of New York discovered the mass grave, the human remains wrapped in layers of linen, at first glance, were mistaken for the bodies of the Coptic monks who lived in the lost monastery of Apa Phoibammon. The monastery was so influential during the Byzantine period that it gave its name to the whole area.[36] Three years later, in 1926, upon closer investigation, the previous evaluation was abandoned in favour of a more intriguing assumption. There were three main factors for reconsidering the previous hasty conclusions concerning the dead bodies.

First, the archaeologists found several linen fabric fragments inscribed with at least twenty-three hieratic surnames within the wrapping of the corpses. According to Herbert Winlock, these anthroponyms were characteristic of the Theban region and dated back to the Eleventh Dynasty.[37]

Second, some fragments of arrows, wooden bows and leather wrist guards (used to protect the inner forearm of the archers against whiplash from the bowstring) were discovered in the tomb, along with the human remains.[38]

Third, the bodies showed several lesions indicating they suffered a violent death in a fight: fifteen individuals had injuries inflicted by arrows, several skulls bore severe traumata caused by the impact of heavy projectiles probably thrown from above and, in some cases, blunt force traumata possibly due to *coupe-de-grâce* blows from a mace.

Moreover, it appears that at least six corpses had been exposed to carrion birds, indicating that they had been left in the open for some time before the burial.[39]

For these reasons, tomb MMA 507 can be considered halfway between a mass grave and a case for battlefield archaeology. The archaeological and anthropological data gathered by the archaeologists allowed them to conclude that the remains belonged to a group of soldiers killed during an attack on a fortified place. Inside the tomb, however, there was no trace of protective clothing and shields. Nevertheless, the thick mass of twisted and greased locks worn by several soldiers would probably provide some protection for the skull. Furthermore, it has also been noted that one of the soldiers had fixed additional artificial locks to his hair, perhaps to provide added protection for his head.[40]

Who were these men? The above-mentioned pieces of evidence and the short distance between the tomb MMA 507 and the mortuary temple of King Nebhepetre Mentuhotep II (c. 2009–1959 BCE) led Herbert Winlock to hypothesize that the remains belonged to some of the soldiers who fought for the Theban sovereign and died during the final stage of the war between the Tenth (Heracleopolitain) and the Eleventh (Theban) Dynasty. Thus, in order to honour their courage in the face of death, the fallen soldiers would have been buried in a tomb overlooking the mortuary monument of their master.[41]

More recently, Sydney Aufrère basically confirmed Winlock's interpretation, corroborating his assumption with the supporting comparison between the arrowheads from tomb MMA 507 and some similar projectiles found in the mortuary complex of Nebhepetre Mentuhotep II.[42] Aufrère also stated that the decoration of the mortuary temple of Nebhepetre Mentuhotep II at Deir el-Bahari should be taken into consideration to shed some light on the collective burial. Among the several relief fragments from the lower colonnade of the temple (PM II, 383–384),[43] a painted limestone fragment, currently kept in the British Museum (EA 732), has especially attracted his attention. This fragment was part of a larger scene showing the siege of a fortified place. On the right, a soldier is depicted climbing to the top of a rampart with the aid of a ladder. On the left, a portion of the rampart is barely visible. In the space between the wall and the ladder, four men pierced by several arrows are falling to the ground from above.[44] According to Sidney Aufrère, this scene would represent the turning point of the war between the Tenth and the Eleventh

Dynasty – probably the siege of Heracleopolis – and the four bodies stuck with arrows would belong to Egyptian soldiers of the army of Nebhepetre Mentuhotep II due to the dark colour of their skin.⁴⁵ Given the extreme rarity of Egyptian military casualties in Pharaonic war scenes, the French scholar has therefore deduced that the relief must necessarily show an episode in an internecine conflict between Egyptians.

It should also be mentioned that the scene from the mortuary temple of Nebhepetre Mentuhotep II has often been related to the well-known contemporary depiction of siege, painted in the *saff* tomb TT 386 of the general Intef 𓀁𓏏𓆑 (PN I, 34, 5) at Thebes (PM I, 437).⁴⁶ There is, without doubt, a rough similarity between the two siege representations but, unlike the relief fragment from the temple of Nebhepetre Mentuhotep II, the scene from the tomb of Intef depicts only corpses of foreign warriors, which have been indeed interpreted as Asiatic men on the basis on their pale skin and their clothing.

We may, therefore, wonder if it is reasonable to link the human remains from the tomb MMA 507 and the siege evoked by the relief fragment from the mortuary temple of Nebhepetre Mentuhotep II. There is no consensus among researchers regarding this issue. Carola Vogel opposed the theory proposed by Herbert Winlock and supported by Sydney Aufrère, putting in question the date attributed to the human remains and the items related to them. Nevertheless, she cannot formulate an alternative hypothesis.⁴⁷

2

Reception and Diffusion of Personal Protections during the Eighteenth Dynasty

As already mentioned in the introductory pages of this book, a new form of warfare emerged in Egypt following the war against the Hyksos and the Early New Kingdom Asiatic campaigns. At the root of that innovative way of fighting was the two-wheeled horse-drawn chariot, the earliest evidence of which dates to the Seventeenth Dynasty.[1] The first inscription mentioning a chariot, written on the so-called second Kamose Stele, dates, as the name suggests, to the reign of Kamose (c. ?–1540 BCE), the last king of the Seventeenth Dynasty, and refers to a chariot used by the Hyksos.[2] The biographical inscription of Ahmose, son of Abana, of a slightly later date, reports on an Egyptian chariot.[3] From then, the two-wheeled chariot would have played a leading role in the Egyptian army for over half a millennium.[4] This true military revolution,[5] which dominated the Western Asian battlefields until the beginning of the first millennium BCE,[6] led almost simultaneously to the adoption of the composite bow.[7] From a tactical point of view, chariots were used as mobile archery platforms, and the composite bow established itself as the principal weapon aboard the chariot. Compared to the traditional self-bow, the composite bow was much more powerful, had a higher rate of fire and gave the archer a more precise aim, even and especially in the 'box' cab of a chariot moving at high speed. To this must be added the emergence of new types of edged weapons, which became common in the New Kingdom as well as the Late Bronze Age Near East: the sickle-shaped sword[8] and the dagger with blade and handle cast together in a single piece.[9]

It was in this context that the need for a combination of protective pieces covering the most vulnerable parts of a warrior's body and giving the advantage of allowing the hands to be free to wield weapons and handle chariots became compelling. The head, thorax, abdomen and groin had all been recognized logically as vital areas of the human body ever since antiquity, so they had to be protected to minimize the risk of debilitating injuries. We can therefore assume that armour provided the best answer to the demand for the protection of charioteers in war.[10]

Most scholars agree that Hurrian-speaking peoples played a fundamental role in the spread of horsemanship and the use of drawn two-wheeled vehicles across the Near East during the second millennium BCE.[11] The famous Hittite text concerning the instructions for the training of chariot horses, purportedly written by the 'master horse trainer from the land of Mitanni' Kikkuli around the mid-fourteenth century BCE, is considered the literary manifesto of the celebrated Hurrian equestrian skills.[12] The members of the Mitannian military elite designated as *Maryannu*,[13] the martial prowess of which was undoubtedly the cornerstone of the expansion of the Mitanni Empire in Northern Mesopotamia during the third quarter of the second millennium BCE,[14] embodied the Hurrian in-depth knowledge of equine breeding and charioteering skills. It even seems that the reputation of the *Maryannu* led Egyptian and Western Asian chariot warriors not only to adopt their equipment and fighting techniques but also to emulate their mode of dress and hairstyle.[15]

Moreover, the development during the Late Bronze Age and the spread throughout the entire Near Eastern region of a new type of corslet composed of rows of overlapping bronze or leather scales are generally attributed to Hurrian-speaking peoples of the Kingdom of Mitanni.[16] As we will see later, even the Egyptian, Hittite and Semitic words for corslet probably derive from the Hurrian term *šarianni/šariyanni* 'body armour'.[17] At the same time, gorgets and conical helmets raised from a single sheet of metal or constituted of several scales are most likely attributable to the same military tradition.

On the other hand, one can imagine the physical toll that the personal protective equipment took on its wearer. In his study *The Western Way of War*, Victor Davis Hanson has shown how demanding it was to wear heavy body armour for a fifth-century BCE hoplite. Bronze helmets, greaves and breastplates were heavy, overheated quickly and imposed limitations on mobility. Moreover, the helmets dangerously affected the peripheral vision and hearing ability of the wearer.[18] The lived experience of armoured Late Bronze charioteers was probably not very different.

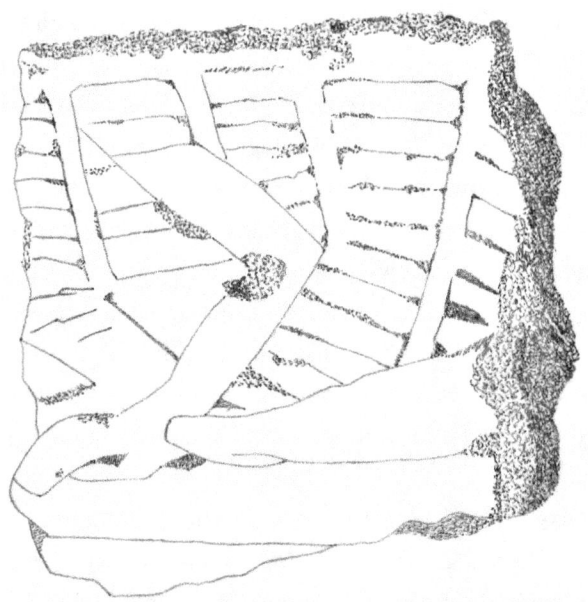

Figure 2.1 Relief fragment ATP 09069 from the mortuary temple of Ahmose I at Abydos, possibly depicting a scale corslet. Drawing by A. M. Pollastrini.

How did this innovative technology reach Egypt at the beginning of the New Kingdom? This is a key question that demands a long and comprehensive answer. We will, therefore, try to detect the plausible channels of supply in light of the archaeological and textual evidence.

2.1 Helmets and armour as ethnic stereotype

Above all, we need to turn our attention to some fragments belonging to the royal narrative reliefs dating to the early Eighteenth Dynasty in order to detect the earliest Egyptian evidence of helmets and armour.

In the last decade of the twentieth century, about fifty painted fragments of relief were discovered in the mortuary complex of King Ahmose I (c. 1539– 1515 BCE) at Abydos (PM V, 92) by Stephen P. Harvey, director of the Ahmose and Tetisheri Project. Several fragments from the eastern side of the inner court of the mortuary temple can be ascribed to a large composition probably depicting the main events which led to the expulsion of the Hyksos from

Avaris and the reunification of the country under the reign of Ahmose I, founder of the Eighteenth Dynasty.[19] They are characterized by high-raised reliefs carved in limestone and painted in bright pastel colours. Among them, the unpublished fragment ATP 09069 (Figure 2.1; appx. III) shows a man (a Hyksos chieftain?) prostrating himself at the feet of the king (Ahmose I?), which appears larger than the other figure, in typical hieratic fashion. Although badly damaged, the relief shows what seems to be a scale or lamellar armour. The bowed figure wears a short-sleeved tunic apparently made of rows of rectangular elements, which look like lamellae or scales laced together in horizontal rows. If this were true, we would be in the presence of the earliest evidence of that type of body armour, which became so important in the Late Bronze Age warfare.

Another large historical narrative relief originally decorated the north wall of the court of the 𓉗𓊚𓋹 *Ḥwt šspt ʿnḫ*, 'Mansion "Receiver of Life"', the dismantled memorial temple of Thutmose II (c. 1482–1480 BCE) in Qurna (PM II, 456).[20] In 1926, Bernard Bruyère discovered a group of fifteen relief fragments showing battle scenes with chariots, archers and dead bodies.[21] The relief was probably released during the reign of Hatshepsut (c. 1479–1458 BCE) or Thutmose III (c. 1479–1425 BCE),[22] in order to commemorate the triumph of Thutmose II over Asiatic foes, perhaps in the course of the punitive expedition against the Shasu raiders in Southern Palestine,[23] or during a relatively unknown military campaign against the kingdom of Mitanni along the Euphrates River.[24] Among this set of block fragments, two, in particular, deserve our attention because of their peculiarities (appx. I).

One of the two fragments clearly shows the death of a bearded Asiatic warrior. In the top left-hand corner is a small part of a plumed helmet, apparently belonging to him. The other fragment shows an Asiatic warrior defeated and wounded by an arrow. Again, a helmet fragment can be seen near the back of a warrior on horseback carrying a quiver.

As far as one can see, both headdresses were conical in shape. Their pointed conical skulls, which seem to be raised from a sheet of metal, are surmounted with an ambiguous inverted v-shaped element (a reinforcing plate? A crest?) and a plume. Neither helmet is worn, but both are part of the debris left on the battlefield by the defeated Asiatics.

The aforementioned examples led us to believe that the armour components have not been represented in the reliefs without a reason or explanation. Iconographic motifs often establish the character of certain groups. Indeed, we suppose that helmets and corslets act as visual representations of stereotypes of Asiatic peoples, especially the northern men from Mitanni.

To this must be added that, from the Eighteenth Dynasty, the presence of helmets lying on the field between the corpses of the defeated enemies become a recurring *topos* within the canon of royal battle scenes – in addition to a range of propagandistic iconographical themes[25] – intended to represent the unconditional surrender of foes, as well as the complete defencelessness of foreigners in the face of Egyptian military power. Since the foreigners living beyond the boundaries of Egypt were considered part of the forces of chaos, the potential threat they posed had to be neutralized by the Pharaoh, who was charged with the responsibility of preserving god-given order. But why choose armour components? In this respect, I believe that the Egyptians arguably perceived personal protective equipment as a distinctive feature of their Asiatic foes, which played a central role in the early process of spreading new military technology from Western Asia to Egypt.

With this in mind, the decorative programme of the chariot body (CGC 46097) found in the Theban tomb (KV 34)[26] of Thutmose IV (*c.* 1400–1390 BCE) must be considered the most evident testimony of the ideological aspects outlined above.[27]

On the exterior right side of the chariot body (Figure 2.2; appx. I, III),[28] the Pharaoh is shown with the Theban war god Montu, firing arrows from his chariot into a group of chariot warriors, some of whom can be identified as North Syrian *Maryannu* by their distinctive physical features (bald or shaved head and pointed beard).[29] In the tangle of fleeing Asiatic war chariots, three armoured warriors are clearly recognizable; one of these, still in control of his chariot, despite being wounded by an arrow, is also wearing a scale helmet and an armoured throat-piece. Furthermore, three helmets are lying on the ground among the battlefield debris.

On the exterior left side (Figure 2.3; appx. I, III),[30] Thutmose IV, standing on his chariot, is shown in the act of striking two Asiatic warriors with a piercing axe that was effective against armoured enemies. This type of weapon

Figure 2.2 Right side of the chariot of Thutmose IV (CGC 46097). Carter, H. and Newberry, P. E., *The Tomb of Thoutmôsis IV*, pl. X.

Figure 2.3 Left side of the chariot of Thutmose IV (CGC 46097). Carter, H. and Newberry, P. E., *The Tomb of Thoutmôsis IV*, pl. XI.

was commonly used during the Eighteenth Dynasty.[31] Again, the foes are seen confusedly retreating in front of the Pharaoh: two among them – one on a chariot and one fallen on the field – are wearing helmets. Moreover, the ground is strewn with abandoned weapons: daggers, shields, quivers and four helmets.

The broad range of armour components minutely carved on the wooden panels of the chariot body deserves further discussion. Timothy Kendall has thoroughly analyzed the different textures and details marking the surface of each helmet in order to identify a correlation between the visual representations and the broad body of specialized terms occurring within the Hurrian period texts recovered from Nuzi (modern Yorghan Tepe, Iraq).[32] Kendall has pointed out the striking similarity between the *gurpisu sippari kurṣimētu* 'bronze scale-covered helmet', described in Nuzi texts, and the helmets made of elongated scales, depicted on Thutmose IV's chariot.

As noted above, one of the helmets belonging to this typology is worn, together with a scale body armour and a studded gorget, by an Asiatic charioteer depicted on the right side of the chariot. Since this carved human figure is generally seen as the most accurate depiction of a Syro-Canaanite chariot warrior of the Late Bronze Age, it often appears in scientific publications concerning Late Bronze and Early Iron Age warfare.[33] This representation is also well-known for the type of injury sustained by the chariot warrior. One of the arrows fired by the Pharaoh pierced the weak point of the armour where the armoured sleeve connects to the corslet, wounding the axillary region between the shoulder girdle and the thorax. The areas that would be hard to encase in the different components of body armour without restricting mobility were understandably the most vulnerable.[34]

Other examples from the Theban region reveal the ideological function of Asiatic helmets and armour in the Eighteenth Dynasty battle scenes. Several relief fragments carved with military scenes have been related to the decoration of the 𓉗𓊹𓏏 *Ḥwt Nb-Khpr.w-Rꜥ m Wꜥst*, 'Mansion of Nebkheperure at Thebes',[35] Tutankhmun's (*c.* ?–1324 BCE) memorial temple built of sandstone blocks taken from Akhenaten's Karnak buildings during the Post-Amarna Period.[36] Unfortunately for now, the function performed by the building and its exact site remains largely obscure.[37] The structure was then demolished under the reign of Horemheb (*c.* 1319–1292),[38] and many of its blocks were reused in the Second Pylon of the Precinct of Amun-Re at Karnak, in the area of Luxor temple and at Medamud as well.[39]

Although the 'Mansion of Nebkheperure at Thebes' was already reduced to countless fragments in ancient times, we have an outstanding range of reconstruction drawings concerning the general aspect of the structure, its decorative programme and the corpus of inscriptions, in particular through the efforts of the researchers of the Centre franco-égyptien d'étude des temples de Karnak (CFEETK) and the Oriental Institute of the University of Chicago.[40] According to these reconstruction attempts, the decorative programme of the temple included scenes commemorating military campaigns against Asiatic and Nubian foes, which represents a relevant antecedent of the Ramesside battle scene tradition.[41]

However, even assuming that the battle scenes do not just have a propagandistic function, it is not easy to prove their historical actuality.

Concerning the Asiatic battle scene, scholars suggest that it depicted a significant clash fought during that phase of the long struggle between the Hittite kingdom and Egypt, which is commonly referred to as the Second Syrian War or the Hurrian War.[42] The fact nevertheless remains that the Second Syrian War was basically a decisive victory of the Hittite king Suppiluliuma (c. 1370–1342 BCE) and a disaster for Egypt and the kingdom of Mitanni. In light of this consideration, is it reasonable to assume that a military defeat became the subject matter of a piece of propaganda art? The answer seems obvious, but the fragmented and, in some way, obscure Hittite and Egyptian written sources leave this question open.[43]

Regarding the Nubian campaign, it should be noted that there is not enough evidence to prove its historicity.[44]

Among the fragments of the Asiatic battle scene, four small sandstone blocks currently kept in the stores of Luxor Temple show plumed helmets lying on the ground among the battlefield debris. According to W. Raymond Johnson, they were unquestionably worn by the defeated Asiatic charioteers who are often depicted lying on the ground near them.[45] They are as follows:

- Block ATP F 836-8 was found reused as filling in the Second Pylon of the Precinct of Amun-Re at Karnak (PM II, 40–41). This raised relief fragment (52 x 26 x 24 centimetres) shows left to right two corpses of

Asiatic warriors lying on the ground, a quiver, a helmet conical in shape (appx. I) and the legs of an advancing Egyptian foot-soldier;[46]
- Block ATP F 882-1 was found reused as filling in the Second Pylon of the Precinct of Amun-Re at Karnak (PM II, 40–41). The raised relief fragment (52 x 26 x 24 centimetres) depicts the legs of three teams of galloping chariot horses and the six-spoked wheels of the vehicles they pull. The wheels of the chariots are running over the bodies of two fallen Canaanite warriors. On the left, a helmet distinguished by a pointed conical skull lays on the ground (appx. I). The headdress is surmounted by an inverted v-shaped element (a reinforcing plate? A crest?) and a feather.[47]
- Block E.S. #832 was found reused in the foundation of a medieval (or modern) building erected in the area in front of the Pylon of the Luxor Temple (PM II, 339).[48] The decoration of the relief fragment (20 x 27.5 x 24.5) is divided into two registers. In the upper register, one can see the legs of at least two enemies swept away by an Egyptian war chariot. A tasselled, conical helmet lies on the ground where its owner had dropped it (appx. I).[49]
- Block ATP 01 has been discovered in an undetermined location within the Luxor Temple complex. The relief is divided into two registers separated by a broad line. The upper register shows the lower part of the legs of a team of chariot horses pulling a chariot behind it. In the lower register, a tasselled, conical helmet appears between the corpses of Asiatic foes lying on the ground (appx. I).[50]

As far as we know, the $\overline{}_{ooo}$ *Mn-mn.w*, 'Most Established of Monuments' the funerary temple of King Ay (*c.* 1323–1320 BCE), subsequently usurped by Horemheb,[51] suffered the same fate as the 'Mansion of Nebkheperure at Thebes'. The structure, which originally stood on the West Bank of the Nile, not far from the site where Ramesses III's mortuary temple was to be built (PM II, 457–460),[52] was possibly dismantled during the Twentieth Dynasty. Again, the extracted stone blocks have been widely reused in the construction of monumental buildings in the Theban region. W. R. Johnson has identified at least seventeen relief fragments quarried and transported from the 'Most Established of Monuments' and reused in the Temple of Khonsu at Karnak.[53] Most stone blocks depict details of Asiatic and Nubian battle scenes. In the

Figure 2.4 Relief fragment from the mortuary temple of Horemheb reused in the forecourt of the temple of Khonsu at Karnak. The Epigraphic Survey, *Temple of Khonsu*. Vol. I, pl. 61. (Courtesy of the ISAC, Chicago.)

group of blocks from the mortuary temple of Horemheb reused in the Temple of Khonsu, one is incorporated into the eastern wall of the forecourt (PM II, 230, [21]) decorated with ritual scenes commissioned by the High Priest of Amun, Herihor (*c*. 1080–1074 BCE).[54] On this fragment (Figure 2.4 and Figure 2.5, appx. I),[55] one can see an Egyptian chariot – which will be discussed later – running through the battlefield covered with corpses of Asiatic warriors. Two conical helmets lie on the ground next to the dead bodies of their owner.

Although equipment designed to protect the wearer's body gradually became more common among Egyptian fighters, at least from the late Eighteenth Dynasty, it continued to be charged with an ideological function even during the Nineteenth Dynasty. The foreign enemies associated with protective clothing are invariably Western Asian warriors. Within the large corpus of Ramesside historical-narrative reliefs, this iconographic topos is recalled in three cases. They are as follows:

Personal Protections during the Eighteenth Dynasty 31

Figure 2.5 Relief fragment from the mortuary temple of Horemheb reused in the forecourt of the temple of Khonsu at Karnak. Prisse d'Avennes, A., *Histoire de l'art égyptien*, pl. III. 13.

- The battle scene of Seti I (*c.* 1290–1279 BCE) on the outside of the northern wall (east wing) of the Hypostyle Hall in the Precinct of Amun-Ra at Karnak depicting the Pharaoh attacking the town of Yenoam[56] in Canaan (PM II, 54, [167]).[57] In the middle of the scene (Figure 2.6; appx. II), Seti I stands on his chariot with his right foot on the chariot pole, grasping two defeated Canaanite enemies shown in larger size than the average Asiatic warrior. Their size, combined with the fact that they are the only ones wearing open-face helmets, leads us to surmise that the two men are part of the Canaanite elite. The helmets are characterized by pointed conical skulls which extend downwards at the rear and sides to afford protection for the neck. Moreover, the helmets have little horsehair plumes fitted to the apex of the skull.
- The battle scene of Seti I carved on the outside of the northern wall (west wing) of the Hypostyle Hall in the Precinct of Amun-Ra at Karnak shows the Pharaoh storming the town of Kadesh and subduing the Kingdom of

Figure 2.6 Seti I subdues the town of Yenoam and Lebanon. Precint of Amun-Ra at Karnak, Hypostyle Hall, north wall, outer face. (Detail.) The Epigraphic Survey, *Relief and Inscriptions at Karnak,* Vol. IV, pl. 11. (Courtesy of the ISAC, Chicago.)

Amurru (PM II, 54, [167]).[58] An Asiatic formation is overwhelmed by Seti's charge before the wall of Kadesh (Figure 2.7; appx. II). Among the Syrian warriors struck by the arrows of Seti I,[59] eight wear tight headdresses surmounted by little plumes. The actual nature of these coverings for the head is difficult to understand. They could be combat helmets or charioteer leather caps.

- The scene of the battle of Kadesh carved on the northern wall of the Hypostyle Hall of the Great Temple of Abu Simbel (PM VII, 103, [41]–[42]).[60] In the top left-hand corner of the great relief, Ramesses II is depicted alone on his chariot sweeping away the Hittite charioteers launched against the Egyptian camp defended only by the Royal Guard and the Amon's division.[61] As usual, the battlefield is littered with debris of battle and corpses of defeated enemies (Figure 2.8; appx. II). Close to Ramesses II's chariot, three helmets are lying between the bodies of the fallen Hittites.[62] The three headdresses are stylized and have the appearance of medieval sallets, which provide good protection for the head and the neck.

Finally, to the large corpus of relief sculptures listed above must be added a three-dimensional sculpture, currently kept in the Basel Museum of Ancient Art (Inv.-Nr. LgAe NN65).[63] The small limestone statue (14.7 x 5.8 x 6.2 centimetres), probably dating from the Eighteenth Dynasty, belongs to the

Personal Protections during the Eighteenth Dynasty 33

Figure 2.7 Seti I subdues the town of Kadesh and the land of Amurru. Precinct of Amun-Ra at Karnak, Hypostyle Hall, north wall, outer face. The Epigraphic Survey, *Relief and Inscriptions at Karnak*, Vol. IV, pl. 23. (Courtesy of the ISAC, Chicago.)

Figure 2.8 Hittite warriors killed by Ramesses II during the battle of Kadesh. Great Temple of Abu Simbel, Hypostyle Hall Drawing, north wall. Drawing by A. M. Pollastrini.

well-known corpus of statuettes used in execration rituals intended to grant protection against the evil and destroy enemies.[64] It depicts a kneeling, bound foreign captive featuring some of the distinctive physical attributes that were characteristic of the Western Asiatic men in Egyptian art (Figure 2.9).[65] The subjugated human figure indeed has a pointed beard and long hair, and wears a plain kilt and a peculiar headdress.

Figure 2.9 Statuette of Western Asiatic (?) kneeling captive. Basel Museum of Ancient Art (Inv.-Nr. LgAe NN65). Di Natale, A. and Basile, C., eds, *Atti del XVIII Convegno di Egittologia e Papirologia, Siracusa, 20–23 Settembre 2018*, 149. (Courtesy of Museo del Papiro 'Corrado Basile', Siracusa.)

The distinctive shape of the latter, in particular, draws our attention. According to André Wiese, curator of the Basel Museum of Ancient Art, the impressive headdress depicted on the foreign prisoner is a conical cap, which, however, has no convincing comparison amongst the several New Kingdom representations of Levantine men. In fact, to our knowledge, there is only one other sculpted portrayal of an Asiatic man with his head covered. We are talking about a calcite head, probably dating from the Eighteenth Dynasty and currently preserved at the Royal Museum of Art and History of Brussels

(Inv-Nr. E.6421),[66] which depicts a bearded man wearing a baggy cap with the apex bent over (now missing). As can easily be seen, there are no similarities between the two headdresses. We, in contrast to Wiese, believe that the headgear of the Basel statuette has many more affinities with the contemporary helmets: it is indeed characterized by a tall conical single-piece skull which extends downwards at the rear and sides to afford protection for the neck and the ears.[67]

2.2 Helmets and armour as prizes of war

War was an opportunity for the Egyptians not only to enhance their knowledge of emerging military technologies, but also a primary source of weapon supply through the collection of the spoils of the enemy on the battlefield.[68] According to Daniel H. Lew, the term 'war booty' 'is limited to moveable articles on the battlefield and in besieged towns. Private property which may be taken as booty is restricted to arms, munitions, pieces of equipment, horses, military papers, and the like'.[69] War booty was undoubtedly an important source of income for Egyptian soldiers and the state itself, but the relevance of the captured enemy property did not lie only in the economic dimension. Indeed, Mario Liverani argues that the antagonist, who dares to hold firm against the Pharaoh and is defeated, is reduced to 'un'entità passiva da saccheggiare o devastare'.[70] With this in mind, captives, animals, weapons, raw materials and moveable valuable objects in general, in other words, everything captured from the enemy, became, through the filter of the Pharaonic ideology, trophies to be listed in the official texts and, perhaps, displayed publicly.

Body armour was mentioned for the first time[71] in the inventory of booty taken after the capture of the fortified town of Megiddo (modern Tell el-Mesettelim, Israel). The siege occurred after a pitched battle fought under the walls of Megiddo by Thutmose III during his twenty-second regnal year (the first as sole ruler) against a coalition of Canaanite rulers led by the king of Kadesh.[72] The short extract of the *Annals* engraved on the north wall of the passage round the bark shrine of Philippus Arrhidaeus (PM II, 97, [280]) in the Precinct of Amun-Ra at Karnak reads as follows:

Annals of Thutmose III, Sect. I (*Urk* IV, 664, 3–5)
Col. 97

ḥzmn mss nfr n ʿḥꜣ n ḫfty pf 1
ḥzmn mss nfr n ʿḥꜣ n wr n Mk[ti 1]
[. . .] mss n ʿḥꜣ n mšʿ=f ḥẓ 200

'Superior quality combat bronze garment belonging to that enemy (the prince of Kadesh), 1;

Superior quality combat bronze garment belonging to the chief of Megiddo, 1;

Combat [. . .] garment belonging to his ruined army, 200.'

The weapons acquired as war trophies during the battle of Megiddo are also recorded, more briefly, on the Victory Stele of Thutmose III from Gebel Barkal (PM VII, 217, [20]).[73] The stele is now kept in the Museum of Fine Arts of Boston (Accession number 23.733):

Victory Stele of Thutmose III from Gebel Barkal (*Urk* IV, 1235.9)
Line 22

Mss.w=sn nb n ʿḥꜣ pḏ.wt=sn šzr.w=sn ḫʿ.w=sn nb

'All their combat garments, their bows, their arrows and all their weapons.'

Later, in Year 35 of his reign, Thutmose III and the Egyptian army captured another haul of helmets and armour from the battlefield after a clash against a coalition of Mitanni vassals next to the still-unknown Syrian town of Arana.[74] The list of booty was originally engraved on the north wall of the vestibule of the Sixth Pylon at Karnak (PM II, 89, [240]–[244]), but it is currently kept in the Louvre Museum (N 205 (C 51)).[75]

According to the official inventory of booty, Thutmose III personally captured two bronze corslets and an unknown number of bronze helmets:

Annals of Thutmose III, Sect. V (*Urk* IV, 711, 7–8)
Col. 39–40

[𓊵𓏠𓈖𓄚𓏤𓈖]𓍱𓏤𓏤

𓊵𓈖𓇋𓆓[𓍱𓏤𓁶𓏤]𓏤𓏤𓏤

[ḥzmn mss n] ꜥḥꜣ 2
ḥzmn dbn [n tp/ḏꜣḏꜣ] [. . .]
'Bronze combat garment 2;
bronze helmet for the head [. . .]'

On the same occasion, Thutmose III's troops took from the enemy an unknown number of bronze corslets and five bronze helmets as spoils:

Annals of Thutmose III, Sect. V (*Urk* IV, 711, 16–712, 1)
Col. 41

𓊵𓏠𓈖𓄚𓏤𓍱𓏤

𓊵𓈖𓇋𓆓𓍱𓁶𓏤𓏤𓏤𓏤𓏤

ḥzmn mss n ꜥḥꜣ [. . .]
ḥzmn dbn n tp/ḏꜣḏꜣ
'Bronze combat garment [. . .];
bronze helmet for the head 5'

The last-known reference to plundered armour comes from a monumental stele erected by Amenhotep II (c. 1425–1400 BCE) in front of the south face of the Eighth Pylon in the Precinct of Amun-Ra at Karnak (PM II, 177, R).[76] The text of the stele records two Asiatic campaigns of Amenhotep II, namely in his seventh and ninth regnal year respectively.[77] During the campaign of the Year 7 in the Beqaa Valley – possibly the most important military operation conducted by Amenhotep II in Western Asia – the rearguard of the Pharaonic army was attacked by a small force of chariots, while it was crossing of the Orontes river.[78] In the aftermath of this skirmish, Egyptian soldiers captured a prisoner, a pair of horses and a limited quantity of weapons and equipment, including a body armour:

Karnak Stele of Amenhotep II (*Urk* IV, 1311, 15–17)
Line 9

[𓍱𓏤𓄚𓈖𓋴𓏠𓈖𓍱𓏤]

[mry]n 1, ssmt 2, wrrt 1,
mss n ꜥḥꜣ (1), pḏ.t 2, ispt mḥty m [ꜥḥꜣ].w 1,
ḥr 1, mꜣšḳw 1, ḫl sḥn.w 1
'1 Maryannu, 2 horse, 1 chariot,
(1) corslet, 2 bow, 1 quiver full of [arrow]s
1 ḥr, 1 mꜣšḳw, 1 inlaid harness'

2.3 Helmets and armour as tribute or gift

According to the *Annals* of Thutmose III, the Pharaoh was honoured with a constant array of foreign gifts and tributes in a period ranging from the campaign fought in Year 33 until Year 42.[79] Moreover, from Year 38, the annual *inw* coming from Retjenu, the region roughly comprising Canaan and Syria, included a certain number of corslets. Scholars have long investigated the meaning of the term *inw*, giving interpretations that range from 'gift' to 'tribute', depending on the ideological justification applied to them.[80] We will not dwell on this topic here. However, I embrace the concept expressed by Mario Liverani,[81] who considers both the terms 'gift' and 'tribute' as misleading when used according to the modern knowledge of the connections between Egypt and the producing countries. It is probable that *inw* had a more neutral meaning and was used, with propagandistic effect, to encompass tributes and traded goods as well.

The *Annals* reveal that corslets were delivered not only in the 38th but also in the 41st and 42nd years of the reign of Thutmose III. The fact that this type of protective clothing reached Egypt only and exclusively from Retjenu leads us to suppose that during the Thutmoside period, the Syrian area was a significant hub for the supply of armour.

Tribute for Year 38
Annals of Thutmose III, Sect. V (*Urk* IV, 718, 17)
Col. 90

[ḥzmn mss n ꜥḥꜣ 4]1
'[Bronze combat garment 4]1'

Tribute for Year 41

Annals of Thutmose III, Sect. VI (*Urk* IV, 726, 17)

Col. 5

𓎺𓏤:𓌉𓐼𓐛𓂋—𓍱

ḥzmn [mss n ʿḥꜣ]

'Bronze [combat garment]'

Tribute for Year 42

Annals of Thutmose III, Sect. VI (*Urk* IV, 732, 1)

Col. 15

𓎺𓏤:𓌉𓐼𓐛𓂋—𓍱

ḥzmn mss n ʿḥꜣ

'Bronze combat garment'

During the same period, representations of the ceremony of *inw* offerings, which took place on the occasion of the New Year Festival in the presence of the king seated in a kiosk,[82] became popular in the tombs of high officials.[83] In the necropolis of Sheikh Abd el-Qurna on the West Bank at Thebes, six scenes include helmets and coats of scales among the foreign goods offered to the Pharaoh:

- Two sets of three helmets each, arranged on trays borne by two Western Asiatic men (appx. I), are depicted on the northern wall of the traverse room in the 'T-shaped' tomb TT 86 of Menkheperrasoneb 𓊹𓏤𓉐𓏤 (PN I, 150, 15), dating from the reign of Thutmose III (PM I/1, 177, [8]).[84] The helmets are painted yellow perhaps to represent the polished bronze. Red and blue plumes surmount the top of three of them.
- A set of ten plumed helmets arranged in two rows on a tray (appx. I) is carved on the southwest wall of the tomb TT 109 of Min 𓏠 (PN I, 151, 14), dating from the reign of Thutmose III (PM I/1, 226–227).[85]
- Two sets of seven and three tall conical helmets, arranged on trays borne by two Western Asiatic men (appx. I), are depicted on the northern wall of the traverse room of the 'T-shaped' tomb TT 42 of Amenmose 𓇋𓏠𓈖𓄟𓋴 (PN I, 29, 8), who lived during the reigns of Thutmose III and Amenhotep II (PM I/1, 82).[86] The helmets are painted white perhaps to represent an unidentified metal/alloy. Moreover, an inverted v-shaped blue element and a plume surmount some of them.

- A set of ten tall conical helmets, arranged on two superimposed small tables (appx. I) are depicted together with other weapons on the southern wall of the traverse room of the 'T-shaped' tomb TT 100 of Rekhmire 𓏞𓏤𓁹𓂋 (PN I, 226, 9), who lived under the reigns of Thutmose III and Amenhotep II (PM I/2, 209, [7]).[87] The helmets are painted yellow perhaps to represent the polished bronze.

- A set of eight helmets and two scale corslets comprising high neck guards (appx. I) are depicted on the western wall of the traverse room of the 'T-shaped' tomb TT 93 of Qenamun 𓎡𓈖𓇋𓏠𓈖 (PN I, 334, 18), dating from the reign of Amenhotep II (PM I/1, 191, [9]).[88] The conical helmets are painted white possibly to represent an unidentified metal/alloy. The two corslets (appx. II) look like short-sleeve tunics made of yellow scales arranged in neat rows. Each scale is marked with a central rib. Yigael Yadin established that no fewer than 450 scales would have been required to make a body armour of the type depicted on the wall of the tomb of Qenamun.[89] The neck, sleeves and lower edge of the corslets are adorned with a blue and yellow stripe pattern which possibly suggests the presence of an internal lining.

  Unusual high neck guards made of bronze plate complete the two suits of armour. Their flared cylindrical shape has often been related to the bronze throat-guard belonging to the well-known Mycenaean panoply, dating from the mid-fifteenth century BCE (LBII/IIB), which has been uncovered in Chamber Tomb 12 of the Dendra necropolis in the Argolid, Greece.[90]

  Finally, it should be noted that this visual representation has given rise to confusion among researchers. According to Thomas Hulit,[91] another set of two corslets, very similar to those represented in the tomb of Qenamun, was depicted in the Eighteenth Dynasty tomb of Paimose 𓊪𓄿𓇋𓄿𓄓𓀀 in the necropolis of Dra Abu el-Naga on the West Bank of the Nile at Thebes (PM I/1, 451). Due to the scarcity of evidence concerning the tomb of Paimose,[92] determining the accuracy of the previous assertion is a challenging task. Nevertheless, Hulit, relying on plate 64 reproduced in Karl Richard Lepsius *Denkmäler aus Ägypten und Äthiopien, Tafelwerke Abt.* III.[93] describes the two corslets as follows: 'The coats of armour [...] are slightly stylized and do not accurately reflect the actual construction of the coats of armour'.[94]

However, upon closer analysis, we can notice that the above-mentioned plate actually depicts a portion of the pictorial decoration of the traverse room of the tomb of Qenamun. How can this strange misunderstanding be explained? It seems plausible that Hulit has mistaken the tomb of Qenamun, which is labelled as 'Theben. Abd el Qurna. Grab 13' in *Denkmäler aus Ägypten und Äthiopien*, with the tomb of Paimose, which is numbered A.13 in *Topographical Bibliography* (Porter & Moss). Unfortunately, scholars have not stopped repeating this old misinterpretation in their works.[95]

- A set of five rounded helmets (appx. I), arranged on a small table, is depicted on the west wall of the traverse room of the 'T-shaped' tomb TT 96A of Sennufer ⬇ (PN I, 309, 5), dating from the reign of Amenhotep II (PM I/1, 198, [6]).[96] A photograph recently taken by the *Mission archéologique dans la Nécropole thébaine* (*MANT*) of the Free University of Brussels shows the helmets painted in pink and surrounded by red lines.

Furthermore, clay tablet EA 22 from the Amarna archive provides a reference to protective gear offered as gifts.[97] The 'Akkadian' text describes the list of gifts sent to Egypt by the Mitanni king Tushratta to celebrate the marriage of his daughter, Tadu-hepa, to Amunhotep III. These gifts include a bronze corslet, a bronze helmet, a leather corslet, a bronze helmet for a *sarku*-soldier, items of protection for horses reinforced with bronze rings, and two bronze head-protectors for horses.[98]

2.4 The spread of the protective clothing among the men serving in the Egyptian army

The battle reliefs carved on the late Eighteenth Dynasty's monuments do not just depict the Western Asiatic enemies of the Pharaoh wearing new-concept helmets and body armour. They are also proof of the spread of this military technology through the territories of the 'Egyptian Empire'. Nevertheless, access to the personal protective equipment was not uniformly distributed among Egyptian soldiers. The iconographic sources indeed show that only troops serving in chariotry were provided with helmets to some degree.

Two fragments belonging to the Asiatic battle relief from the 'Mansion of Nebkheperure at Thebes' – M. 4487 found at Medamud (PM V, 159),[99] and E.S.

#832 found reused in the foundation of a medieval (or modern) building erected in the area in front of the Pylon of the Luxor Temple (PM II, 339)[100] – show the top of several conical helmets, worn by Egyptian charioteers (appx. I).

Another relief fragment related to the decorative programme of the memorial building of Tutankhamun in Karnak (Cairo Museum Reg. Num. 13940, TN 8/6/24/7)[101] depicts an Egyptian charioteer equipped with a bow and a helmet in a parade along with other troops following a military expedition in Nubia (appx. I). The sandstone block reused as filling inside the Second Pylon of the Precinct of Amun-Re at Karnak (PM II, 40) was a portion of a larger composition probably mirroring the Asiatic battle scene Egyptian mentioned above.[102]

The iconographic sources connected to Horemheb's service in the army and his subsequent accession to the Pharaonic throne best represent the appearance of the late Eighteenth Dynasty charioteers. The limestone block MCA-EGI-EG 1889 kept in the Museo Civico Archeologico of Bologna[103] was originally part of the decoration of the funerary complex that Horemheb had built in the Saqqara necropolis when he was still a high official (PM III²/2, 567).[104] The relief fragment, mainly known for the presence of a rare depiction of a horseman,[105] shows a camp scene with a group of six men, probably charioteers, who are carrying a heavy load. These soldiers wear leather loincloths over their cloth garments, which during the New Kingdom had become a distinctive part of military apparel.[106] What catches the eye is that only three men among them wear helmets over their wigs. How can this iconographic detail be interpreted?

To respond to this question, we should once again consider the above-mentioned block (Figure 2.4 and 2.5) transported from the mortuary temple of Horemheb and reused in the Temple of Khonsu at Karnak. Most of the surface of the relief fragment is occupied by a pictorial representation of an Egyptian crew standing on the platform of a war chariot: a mounted bowman equipped with a helmet and a driver bearing a shield.

It should be specified that the organization of the fighting branch of the Egyptian chariotry is still unclear. Although the iconographical sources always show crews comprising a chariot driver and a fighter, New Kingdom texts mention three different types of charioteers: *znny* 'chariot warrior' (*Wb* III, 459, 17-18),[107] *kḏn/kṯn* 'chariot driver' (*Wb* IV, 148, 12-15)[108] and *krʿw* 'shield-bearer' (*Wb* V, 59, 12-14). Jean Yoyotte and Jesus López suggested that starting from the Amarna Period, the words *kḏn/ktn* and *krʿw* began to be used to categorize charioteers – collectively referred to as *znny* – into two groups according to their role. By this theory, the *kḏn/ktn* were the higher-ranking veteran chariot drivers, and the *krʿw* were the cadets.[109] Alan Schulman, in turn, claimed that *krʿw* was an honorific title rather than a military rank.[110] Finally, according to Anthony Spalinger, the use of the three terms in Egyptian texts has evolved significantly through the New Kingdom, making it challenging to identify a precise equivalence between the title and the actual role played in the army.[111]

Whatever the real meaning of these titles, it seems that each member of an Egyptian chariot crew was provided with special equipment during field operations. Both the block reused in the Temple of Khonsu at Karnak and the relief fragments from the 'Mansion of Nebkheperure at Thebes' indicate that the warrior specializing in mounted combat was equipped with a composite bow, quiver and helmet, and the charioteer assigned to drive the vehicle usually bore a shield capable of protecting himself and his companion. Furthermore, the soldier wearing the helmet and the bow always stands in the foreground, partially covering the driver, whatever the direction of the chariot. From this recurring iconographic detail, one has the impression that the charioteer responsible for firing the arrows at the enemy had a higher rank than his crewmate.

Thus, in order to answer the previous question, we can suppose that the soldiers carved on the relief kept in the Museo Civico Archeologico of Bologna were the members of three different crews represented as if they were still standing on their chariot: the bowmen (sporting helmets) in the foreground and the drivers in the background.

Even more telling is the fact that the Egyptian charioteers were never depicted wearing corslets in the reliefs analyzed to date, proving perhaps that the use of body armour was limited to the Pharaoh, high-ranking officers and their retainers.

3

Personal Protective Equipment during the Ramesside Period

3.1 Protective clothing in Ramesside war reliefs

Starting from the Nineteenth Dynasty, the monumental battle reliefs carved on walls of the temples erected in the Theban region show an increasing spread of the use of helmets and body armour not only among the charioteers but also among the men who served in the infantry, especially front-line troops. How to explain this process? Perhaps the answer lies in the necessity to face the changes that occurred in military tactics during the Late Bronze Age, following the appeareance in the Eastern Mediterranean of the 'northern' warriors, better known as Sea Peoples. As noted by Robert Drews, the Sea People invaders were so succesful because they were armed and trained to perform irregular warfare tactics, effectively countering the traditional Western Asiatic chariot tactics.[1]

A section of the aforementioned relief of the battle of Kadesh carved on the northern wall of the Hypostyle Hall of the Great Temple of Abu Simbel (PM VII, 103, [41]–[42]) summarizes well the range of coverings available to the soldiers of Ramesses II. We are referring here to the well-known scene depicting the war council chaired by Ramesses II and the brutal interrogation of two captured spies before the sudden assault of the Hittite chariots on the Ra division (Figure 3.1).[2]

The scene is divided into two registers, with the upper one dominated by the seated figure of Ramesses II. Before the Pharaoh, there is a group of dignitaries and guards, including an armoured charioteer (Menna, the personal charioteer of Ramesses II?) holding the reins of the royal chariot. This man wears an open-faced helmet and body armour with short sleeves made from alternate rows of blue, red and yellow scales.

Figure 3.1 The war council chaired by Ramesses II before the battle of Kadesh. Great Temple of Abu Simbel, Hypostyle Hall, north wall. (Detail.) Champollion, J.-F., *Monuments de l'Égypte et de la Nubie: Planches*. Tome I, pl. XXIX.

In the lower register, below the throne of Ramesses II, there are two groups of Egyptian and Sherden[3] royal guards facing each other and, nearby, a third group of guards attending to the interrogation of the foreign spies. All the soldiers depicted on the relief (except the charioteer) sport a type of body armour known only from iconographic sources: a sleeveless corslet designed to protect the torso, apparently made of stuffed and quilted cloth or thin leather, fastened at one side by loops and buttons. The corslet is worn over a kilt with a triangular front piece, probably consisting of layers of linen fabric quilted together. Unfortunately, nothing is known about this kind of body protection, not even its Egyptian name. Nevertheless, judging from the historical reliefs depicting the military campaigns of Ramesses II and Ramesses III, the aforementioned equipment was exclusively provided to retainers in the service of the Pharaoh. Moreover, the Sherden warriors are equipped with the distinctive round shield, the long pointed sword and the round horned helmet surmounted by a central disc.[4]

Moving on to consider the Karnak Temple complex, a cycle of four reliefs carved on the west wall of the so-called 'Cour de la Cachette', between the Great Hypostyle Hall and the Eighth Pylon (PM II, 133, [493])[5] should be taken into account. These four reliefs depict the conquest of three Canaanite walled towns and a pitched battle in a hilly landscape. At an earlier stage, the four scenes have been attributed to Ramesses II but have been successively reassigned to Merenptah (*c.* 1212–1202 BCE) and correlated to the military operations listed in the Triumph-Hymn of the Victory Stele of Merenptah, better known as Israel Stele.[6] Of the three Asiatic fortified towns depicted on the reliefs at Karnak, Ashkelon is the only one that is specifically named. The town is depicted as a fort consisting of two concentric curtain walls. The ramparts are filled with the town's inhabitants raising their hands and burning incense, possibly as a sign of surrender.[7] Among the Egyptian shock troops, who are assaulting Ashkelon, there are four infantrymen wearing helmets, probably to protect their heads from projectiles dropped from the top of the walls.

To what extent soldiers were forced to provide themselves with military equipment at their own expense during the Ramesside period is a fascinating question that is difficult to answer. Contemporary scribal miscellanies offer two different points of view on the same issue. The hieratic papyrus *Anastasi* III

(BM EA 10246,1), dating back to the third regnal year of Merenptah, includes a fictional composition called *The Suffering of an Army Officer* whose main character, an 'unfortunate' young chariot officer, is compelled to give away his property to buy his own chariot (Pap. Anast. III, 6, 7–8).[8] In the literary papyrus *Lansing* (BM EA 9994), written during the Twentieth Dynasty, in turn, weapons and other equipment are stored in the Fortress of Tjaru (modern Tell Hebua, eastern Nile Delta)[9] in preparation for a military campaign in Syria (Pap. Lansing, 9, 10).[10] As Ole Herslund correctly pointed out, however, these compositions cannot be considered the most reliable form of historical record, as they are ideologically connoted, belonging to 'the group of texts in which scribes present a negative view of the army, its members and lifestyle'.[11] Moreover, a well-preserved relief on the exterior north wall of the mortuary temple of Ramesses III (*c*. 1187–1157 BCE) at Medinet Habu on the West Bank of the Nile, opposite Thebes (PM II, 518, [188]) provides a unique representation of the distribution of weapons to the Egyptian soldiers before a military campaign. This event took place in an unknown location before the land and naval battles fought in the Year 8 of the reign of Ramesses III between the Pharaonic army and the Sea Peoples, who were threatening the northern Egyptian borders (Figure 3.2, appx. II).[12] The left portion of the scene is occupied by an orderly line of soldiers, who are going to receive their equipment from two officials. On the right, two scribes are taking note of each piece of

Figure 3.2 Ramesses III distributes weapons to his soldiers before their campaign against the Sea Peoples. Mortuary Temple of Ramesses III at Medinet Habu, exterior wall, outer face, north side. (Detail.) The Epigraphic Survey, *Medinet Habu*, Vol. I, pl. 29. (Courtesy of the ISAC, Chicago.)

equipment delivered. Between the scribes and the officials a great number of weapons are arranged in superimposed rows: helmets resembling hoods in shape with two tassels attached to the top, spears, quivers, *khopesh* sickle-shaped swords and other unidentified objects, which have been interpreted as folded corslets.[13] Above the scribes, a hieroglyphic inscription clarifies that the weapons are issued to all the branches of the army, including foreign mercenaries:[14]

35. [▨▨▨]▨▨ 36. [▨][▨]▨▨ 37. ▨▨▨
38. ▨▨▨ 39. ▨▨▨▨▨▨ 40. ▨▨▨
35. [. . . . .] [pr] (?) 36. [. . . . . .] [r]dit 37. ḥʿw n mšʿ 38. nt ḥtry 39. n pḏ.t Šrdn.w 40. Nḥsy.w

'[. . . . .] [gi]ving the weapons to the infantry, to the chariotry, to the unit of Sherden and Nubians'

Helmets similar to those shown in the previous relief are worn by the Egyptian soldiers in several other battle scenes carved in the mortuary temple of Ramesses III at Medinet Habu:

- The relief (Figure 3.3) carved on the outer face of the north exterior wall depicting the major battle fought in the land of ▨▨▨ Ḏꜣh(y) 'Djahy'[15] during Year 8 of the reign of Ramesses III between the Egyptian army and the Sea Peoples (PM II, 518, [188]).[16]
- The relief (Figure 3.4) carved on the outer face of the north exterior wall depicting the sea battle between the Egyptian fleet and the Sea Peoples that occurred at the mouth of the Nile somewhere in the eastern Delta in the Year 8 of the reign of Ramesses III (PM II, 518, (188)–(189)).[17] In addition to the helmets, the soldiers serving on the Egyptian ships wear short-sleeve corslets. Both helmets and corslets are characterized by a striped pattern, which could represents superimposed rows of scales or quilted layers of fabric.[18]
- Two reliefs carved on the outer face of the north exterior wall between the First and Second Pylons (PM II, 520–521, [181])[19] and on the east wall of the First Court (PM II, 492, [63])[20] respectively, depicting from two different perspectives the Second Libyan campaign of Ramesses III (Year 11) (Figure 3.5 and 3.6; appx. II).

Figure 3.3 The land battle of Djahy between Ramesses III and the Sea Peoples. Mortuary Temple of Ramesses III at Medinet Habu, exterior wall, outer face, north side. (Detail.) The Epigraphic Survey, *Medinet Habu*, Vol. I, pl. 33, 34. (Courtesy of the ISAC, Chicago.)

Figure 3.4 The naval battle between the Egyptian fleet and the Sea Peoples. Mortuary Temple of Ramesses III at Medinet Habu, exterior wall, outer face, north side. (Detail.) The Epigraphic Survey, *Medinet Habu*, Vol. I, pl. 38, 39. (Courtesy of the ISAC, Chicago.)

Personal Protective Equipment in the Ramesside Period 51

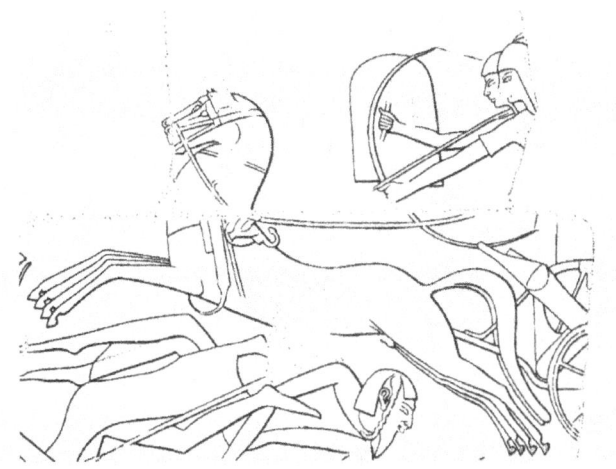

Figure 3.5 Ramesses III pursues the fleeing Libyans during the Second Libyan War (Year 11). Mortuary Temple of Ramesses III at Medinet Habu, exterior wall, outer face, north side, between pylons. (Detail.) The Epigraphic Survey, *Medinet Habu*, Vol. II, pl. 70. (Courtesy of the ISAC, Chicago.)

Figure 3.6 Ramesses III defeats the Libyans during the Second Libyan War (Year 11). Mortuary Temple of Ramesses III at Medinet Habu, first court, east wall. (Detail.) The Epigraphic Survey, *Medinet Habu*, Vol. II, pl. 72.

- The undated relief (Figure 3.7) carved on the outer face of the north exterior wall between the First and Second Pylons depicting Ramesses III storming a Syrian town (PM II, 520–521, [192]). In this scene, only the personal charioteer of Ramesses III wears the helmet.[21]
- The undated relief (Figure 3.8) carved on the north wall of the First Court depicting an Amorite town besieged by Ramesses III (PM II, 494). In this scene, again, only the personal charioteer of Ramesses III wears an helmet.[22]

A different type of helmet is depicted on the battle relief carved on the upper part of the exterior north wall between the First and Second Pylon of Medinet Habu (PM II, 520)[23]. The scene (Figure 3.9)[24] shows in detail an undated siege of Tunip, which, according to the hieroglyphic legend, is located in the Land of Hatti.[25] A group of four Egyptian archers are standing at the foot of the wall while firing arrows in the direction of the town's defenders. In addition to a long tunic, each archer wears an unusual military headdress in the shape of a tight conical cap with two tassels attached to the top (appx. II). These headdresses compare well in shape to that depicted on a figured *ostrakon* from the village of Deir el-Medina, dating back to the Twentieth Dynasty.[26] The fragment of limestone (13.9 x 11.7 centimetres), actually kept in the Medelhavsmuseet of Stockholm (Inv-Nr 14111), shows a charioteer

Figure 3.7 Ramesses III storms a Syrian town. Mortuary Temple of Ramesses III at Medinet Habu, exterior wall, outer face, north side, between pylons. (Detail.) The Epigraphic Survey, *Medinet Habu,* Vol. II, pl. 90. (Courtesy of the ISAC, Chicago.)

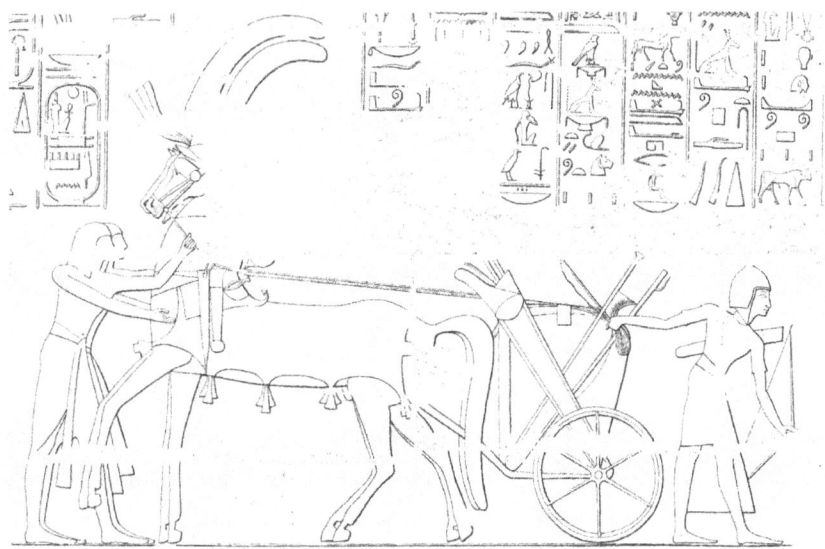

Figure 3.8 A Syrian town besieged by Ramesses III. Mortuary Temple of Ramesses III at Medinet Habu, first court, north side. (Detail.) The Epigraphic Survey, *Medinet Habu*, Vol. II, pl. 94. (Courtesy of the ISAC, Chicago.)

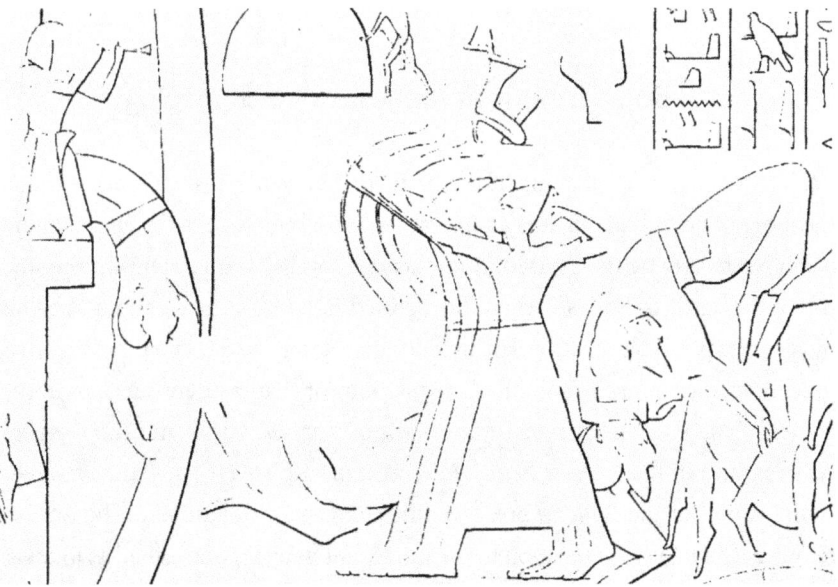

Figure 3.9 The town of Tunip besieged by Ramesses III. Mortuary Temple of Ramesses III at Medinet Habu, exterior wall, outer face, north side, between pylons. (Detail.) The Epigraphic Survey, *Medinet Habu*, Vol. II, pl. 88. (Courtesy of the ISAC, Chicago.)

riding a chariot drawn by two horses, who wears a tall conical helmet (probably made of leather) decorated with stripes and surmounted by two tassels.

In light of this, is it reasonable to suppose that the headdresses depicted in the relief of the siege of Tunip could be charioteer helmets? If this hypothesis were confirmed, it would mean that the archers were actually dismounted charioteers participating in the arrow barrage. Indeed, it is not hard to imagine that the chariots were basically useless during sieges and that charioteers were forced to fight as infantry.

A set of eight similar helmets had been depicted on the right wall of the side room M (Ch) of the hypogeum of Ramesses III (KV 11) at Thebes (PM I/2, 522–523).[27] Even though in that area the painting has disappeared almost entirely, we can get an idea of the original appearance of the helmets through several reproductions of the decorative programme in Room M made in the nineteenth century (Figure 3.12; appx. II).[28] The eight helmets were arranged in two columns and had a red or green painted surface interrupted by black horizontal parallel lines with two tassels hanging from the top. The depiction of a number of objects associated with chariotry on the left wall of Room M might corroborate the hypothesis that headdresses of that type were charioteer helmets.[29]

3.2 Body armour in Ramesside royal ideology

Within the primary accounts of the battle of Kadesh, the so-called *Poem* and the more concise *Bulletin*, we can find the earliest reference to the employment of body armour by the Pharaoh in an actual combat setting. During the first day of the battle of Kadesh, when the fortified Egyptian camp is about to be hit by an unexpected assault by Hittite chariots, Ramesses II, taken by surprise, quickly readies himself for battle. The memory of that incident survives in the copies of the *Poem* preserved on the walls of Karnak, Luxor and Ramesseum temples, and in the *Chester Beatty III* and *Raifé-Sallier III* papyrus, as well as in the copies of the *Bulletin* preserved in Luxor, Ramesseum and Abu Simbel temples. The Pharaoh's reaction to the imminent danger is described as follows:

Poem §§ 76–78 (KRI II, 28, 2–29, 6)
– Karnak, Great Hypostyle Hall, south wall, outer face, PM II², 58, [174][30]

Personal Protective Equipment in the Ramesside Period 55

Col. 17–18

[hieroglyphs]

Ḥm=f [...] šzp.n=f ḥkr.w (n) ʿḥ3 ṯ3(y)=f sw m p3y=f ṯryn sw mi Bʿr m wnw.t=f

'His Majesty [...] he took up his panoply of war, he girded himself with his cuirass – He was like Baal in his moment'

– Karnak, Court between the Ninth and Tenth Pylon, west wall, outer face, PM II², 179, [537][31]

Col. 32–33

[hieroglyphs] *lost*

šzp.n=f ḥkr.w (n) ʿḥ3 ṯ3(y)=f sw [...]

'He took up his panoply of war, he girded himself [...]'

– Luxor, Pylon, outer face, PM II², 304–305, [13]–[14][32]

Col. 22

[hieroglyphs]

Ḥm=f [...] šzp.n=f ḥkr.w (n) ʿḥ3 [............] p3y=f ṯryn sw mi Bʿr m wnw.t=f

'His Majesty [...] he took up his panoply of war, [.........] with his cuirass – He was like Baal in his moment'

– Luxor, Court of Ramesses II, east side, exterior, PM II², 335, [216]–[217][33]

Col. 22

[hieroglyphs]

Ḥm=f [...] šzp.n=f ḥkr.w n ʿḥ3 ṯ3(y)=f sw m p3y=f ṯryn sw mi Bʿr m wnw.t=f

'His Majesty [...] he took up his panoply of war, he girded himself with his cuirass – He was like Baal in his moment'

– Ramesseum, 2nd Pylon, outer face, PM II², 434, [3][34]

Col. 15

[hieroglyphs] *lost*

[...] ṯ3(y)=f sw [...]

'[...] he girded himself [...]'

– Hieratic papyrus *Chester Beatty III*, (BM EA 10683,2)[35]
Ch B1 *verso* 3, 10–11

ṯ3(y)=f sw m p3y=f ṯryn sw mi Bʿr m wnw.t=f

'He girded himself with his cuirass – He was like Baal in his moment'

Ch B2 *verso* 1, 14

[. . .] ṯryn(3) sw mi Bʿr [. . .] =f

'[…] himself with his cuirass – He was like Baal […] his […]'

– Hieratic papyrus *Raifé* (Louvre E 4892) / *Sallier* III (BM EA10181)[36]
Col. 1, 6

Ḥm=f ʿnḫ wḏ3 snb [. . .] šzp.n=f ḫkr.w[37] n ʿḥ3 ṯ3(y)=f p3y=f ṯryn sw m(i) Bʿr m wnw.t=f

'His Majesty, may he live, be prosperous and hale, […] he took up his panoply of war, he girded (himself) with his cuirass – He was like Baal in his moment'

Bulletin §§ 84, 86-87 (KRI II, 119, 7 – 120, 5)
– Luxor, Pylon, east tower, outer face, PM II, 304, [13][38]
Col. 23–24

Ḥm=f [. . .] šzp.n=f ḫkr.w ṯ3(y)=f [. . . . .] =f ṯryn sw mi Swtḫ m 3.t sḫm=f

'His Majesty […] he took up his panoply of war, he girded [……] his cuirass – He was like Seth during his powerful strike'[39]

– Luxor, Court of Ramesses II, east side, exterior, PM II, 335, [218][40]
Col. 20–21

Ḥm=f [. . .] šzp.n=f ḫkr.w ṯ3(y)=f sw m p3y=f ṯryn sw mi Swtḫ m 3.t sḫm=f

'His Majesty […] he took up his panoply of war, he girded himself with his cuirass – He was like Seth during his powerful strike'

– Ramesseum, 1st Pylon, east tower, inner face, PM II, 433, [3][41]
Col. 22-23

Ḥm=f [. . .] šzp.n=f ḫkr.w ṯ3(y)=f [. . . . .]
[ṯr]yn sw mi Bʿr m wnw.t=f
'His Majesty [...] he took up his panoply of war, he girded [...] [cui]rass
– He was like Baal in his moment'

– Great Temple of Abu Simbel, Great Hypostyle Hall, north wall, PM VII, 103-104, [41]–[42][42]
Col. 39

Ḥm=f [. . .] šzp.n=f ḫkr.w ṯ3(y)=f sw m p3y=f ṯryn sw mi Bʿr m wnw.t=f
'His Majesty [...] he took up his panoply of war, he girded himself with his cuirass – He was like Baal in his moment'

Noteworthy is the juxtaposition of Ramesses II wearing his body armour and the martial god Baal, which, according to Hermann te Velde must be 'recognised as a form in which Seth revealed himself'.[43]

In this respect, the close relationship between the members of the Nineteenth Dynasty and the god Seth/Baal should not be forgotten. The worship of Seth during the Rammesside period is well exemplified by the so-called '400 Year Stela' found in Tanis (modern San al-Hagar) in the eastern Delta of the Nile and then moved to the Cairo Museum (JdE 60539).[44] In the lunette, the god Seth, depicted with the appearance of Baal, is worshipped by two figures: Ramesses II and a possible representation of his father, Seti I, when he was still a prominent member of the Egyptian bureaucracy.[45] The purpose of this stele is not clear but it is generally assumed that it refers to the commemoration of a 400th anniversary related to the god Seth. However, it can be said with relative certainty that the stele was intended for the promotion of the worship of Seth through the connection of the Pharaoh and his ancestors to this powerful deity.[46]

But what does the text mean when it says Ramesses II 'was like Seth during his powerful strike'? And what is the 'moment of Baal' mentioned here? Might it refer to the myth about the gods fight against the Sea,[47] mainly known in the Egyptian literature from the so-called *Astarte Papyrus* (Pap. Amherst IX), dated to the fifth regnal year of Amenhotep II?[48]

In answering this question, it is important to take into account the Papyrus Bibliothèque Nationale 202, recently accepted as the first page of the *Astarte Papyrus*. In the text of this fragmentary papyrus, Seth/Baal, the hero of the story described by Thomas Schneider as 'a prototype of the belligerent kingship'[49], is equipped with body armour in a similar manner to Ramesses II at Kadesh. In the light of this supposed similarity between the king and the god, can we assume that the ramesside ideology wanted to combine Seth/Baal fighting the sea with Ramesses II opposing the uncountable waves of Hittite chariots? Although this hypothesis is fascinating, there is currently not enough evidence to support it.

Turning our attention to the iconographical sources, on the inner face of the Second Pylon of the Ramesseum temple at Thebes, there is a relief that could be related to the passages of the *Poem* and the *Bulletin* mentioned above (PM II, 434, [10]).[50] The scene, which has no parallels in New Kingdom iconography, depicts the king equipped with his body armour, charging the Hittite troops on his chariot before the walled city of Kadesh, near the Orontes River. The well-known colour detail of this scene, painted by Prisse d'Avennes,[51] shows in minute detail the coat of scales worn by the Pharaoh: a full-length corslet with short sleeves made from alternate rows of blue and yellow small scales (Figure 3.10).

In addition, it should be noted that some of the dead bodies lying on the battlefield wear flamboyant scale corslets which look very similar to the body armour of Ramesses II. Unfortunately, nowadays, the original colours have almost completely disappeared from the relief.

A piece of further evidence provides a remarkable detail relating to Ramesses connection with his body armour. We refer here to the short rhetorical text associated with the undated reliefs of the siege of Dapur, carved twice, respectively on the entrance wall of the Ramesseum Hypostyle Hall (PM II, 438, [17])[52] and the exterior western wall of the Second Hypostyle Hall of Luxor temple (PM II, 333, [202]–[203]).[53] Not much is known about the actual siege described in the reliefs and related texts, except that it involved the Syrian town of Dapur at an uncertain moment after the eighth regnal year of Ramesses II.[54]

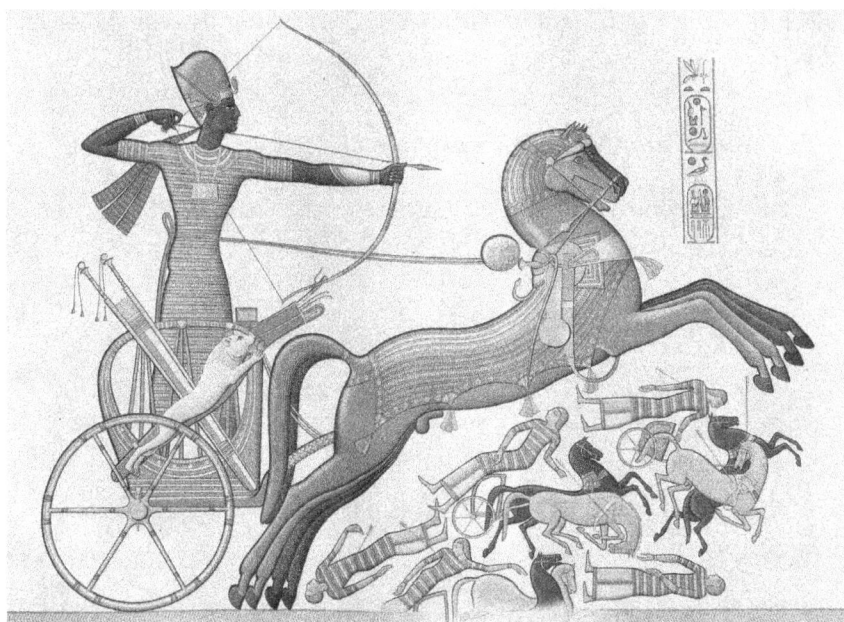

Figure 3.10 Ramesses II on his chariot at the battle of Kadesh, wearing his scale corslet. Ramesseum, Second pylon, inner face. Prisse d'Avennes, A., *Histoire de l'art égyptien*, pl. III. 30.

Siege of the Syrian town of Dapur (KRI II, 175, 3–12)
– Ramesseum, Hypostyle Hall, entrance wall, PM II, 438, [17]
Col. 4–7

4. [...] 𓏛𓈖𓅓𓏤𓇾𓏏𓂝𓊪𓇋𓇌𓆑𓍿𓂋𓇋𓈖

5. [▨ lost ▨] ... 𓊪𓇌𓆑𓏏𓂋𓇋𓈖

6. [▨ lost ▨] ...

7. [▨ lost ▨] ...

4. [. . .] *ir.n ḥm=f tȝ.t pȝy=f tryn*
5. [. . . . .*lqst*. . . . .] *wnw.t 2 iw=f ʿḥʿ ḥr ʿḥȝ pȝ dmi n nȝ ḫrw.w n Ḥt r-ḥȝt pȝy=f mšʿ tȝy=f*

6. [. . . .*lqst*. . . .] [*ṯ*]*ryn ḥr=f ir.n ḥm=f ii r ṯз.t pзy=f ṯryn ʿnn rdi.tw=f ḥr=f iw*

7. [. . . . .*lqst*. . . . .] *n Ḫt nty m w n dmi Twnp m pз tз Nhrnз iw bn pз=f ṯryn ḥr=f*

– Luxor, Second Hypostyle Hall, exterior western wall, PM II, 333, [202]–[203][55]

Col. 4–7

4. [...] ir.n ḥm=f tз.t pзy=f ṯryn(з) r di.tw=f ḥr=f [iw] ḥm=f ʿnḫ wḏз snb wnw.t 2 iw=f

5. ʿḥʿ ḥr ʿḥз pз dmi n nз ḥrw.w n Ḫt r-ḥзt pзy=f mšʿ tзy=f nt ḥtr [iw bn] ṯryn ḥr=f ir.n ḥm=f

6. ii r tз.t pзy=f ṯryn ʿnn rdi.tw=f ḥr=f iw iry=f tзy wnw.t 2 n ʿḥз pз dmi n nз ḥrw.w n Ḫt (nty) m

7. w n dmi Twnp m pз tз Nhrnз iw bn pз=f ṯryn ḥr=f

His Majesty took his cuirass to put it on, after His Majesty, Life, Strength and Health, had spent two hours standing there and attacking the town of the Hittite enemies, in front of his troops and chariotry [without] a cuirass upon him. Now, His Majesty returned to pick up his cuirass to wear it, when he had spent these two hours in [attacking] the town of the Hittite enemies, in the region of the town of Tunip, in the land of Naharina, without any armour on him.

This anecdote has no parallel in royal propagandistic literature. A reaction probably determined by the hot climate becomes, through the prism of the ideology, an act of gallantry witnessed by the whole army. The soldiers would believe in victory because they were under the command and had the guidance of the gallant Pharaoh. Indeed, it is well known that morale wins wars.

Ramesses II shows, therefore, his heroic, almost superhuman, nature, fighting on the front line without the protection of his body armour for two hours, a glorious deed that, as rightly noted by Anthony Spalinger, 'further reinforces our opinion of the king as a doughty war leader'.[56]

According to Claude Obsomer,[57] the relief engraved in the Luxor temple might depict the exact moment reported in the rhetorical text. There, Ramesses II is indeed represented wearing only a short sleeve tunic while shooting an arrow toward the walls of Dapur.

In this context, it must be noted that if we exclude the representation of thirteen colourful coats of scales (Figure 3.11; appx. III) and eight helmets (Figure 3.12; appx. II) on the walls of Room M in the tomb KV 11 of Ramesses III,[58] all the Ramesside sources linked to the interaction between the Pharaoh and his body armour, are concentrated in the reign of Ramesses II. It cannot be determined, in our present state of knowledge, why Ramesses II chose, to a certain extent, to bind his image both iconographically and textually to the body armour. Could this piece of equipment have a special meaning for

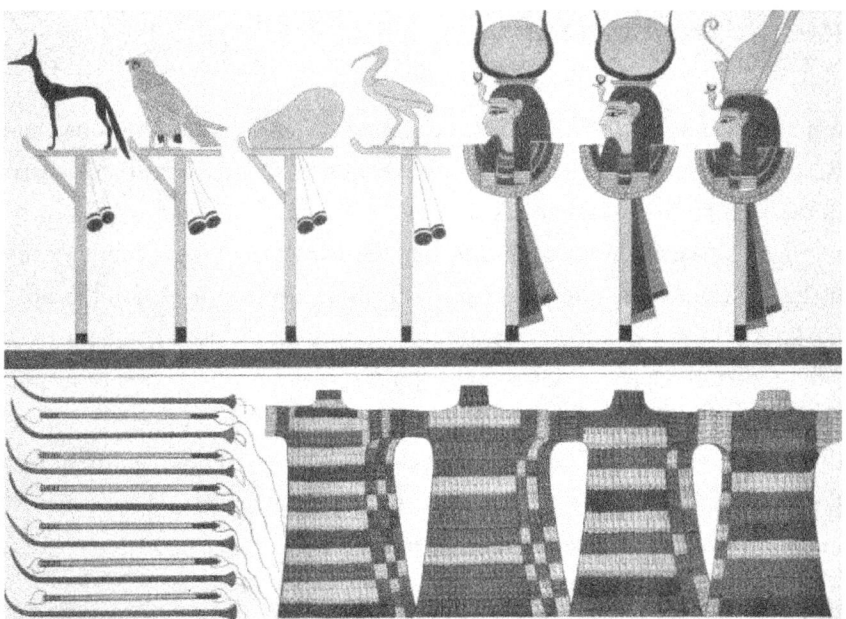

Figure 3.11 Standards and weapons. Tomb of Ramesses III (KV 11), Room M (side chamber Ch), left wall. Champollion, J.-F., *Monuments de l'Égypte et de la Nubie: Planches*. Tome III, pl. CCLXII.

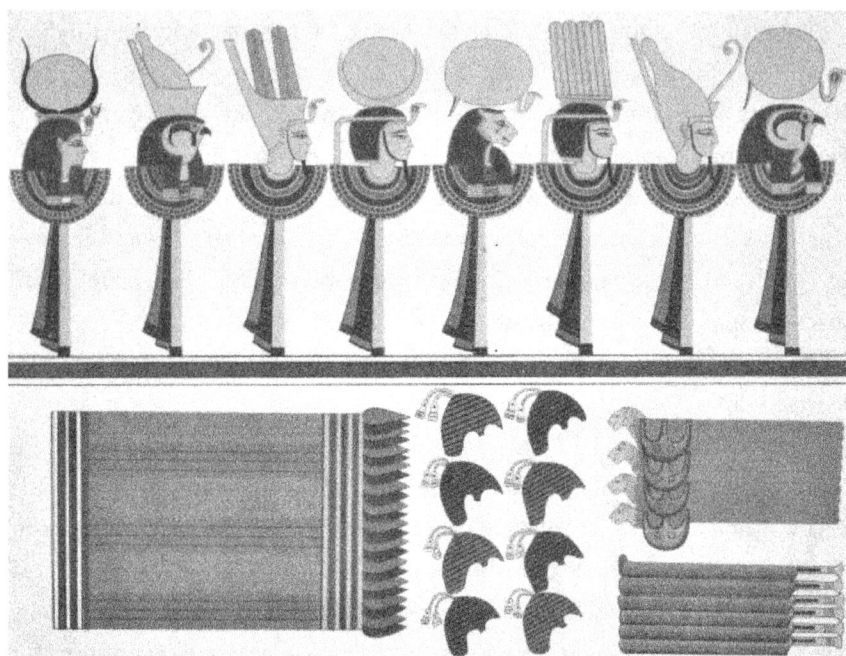

Figure 3.12 Standards and weapons. Tomb of Ramesses III (KV 11), Room M (side chamber Ch), right wall. Champollion, J.-F., *Monuments de l'Égypte et de la Nubie: Planches.* Tome III, pl. CCLXIII.

him? Unfortunately, there are no unequivocal answers to this question. According to sources collected here, two very different aspects seem to coexist in the body armour of Ramesses II.

On the one hand, we can presume that the extract from the accounts of the battle of Kadesh was an integral part of the royal ideology, related to the close relationship between the Ramesside rulers and their dynastic god Seth/Baal, whose warrior aspect is not actually limited to the myth of the Storm-god fighting against the Sea.[59] The official representation of the armoured king on the Second Pylon of the Ramesseum could indeed support this hypothesis. On the other hand, fighting without armour was an act of bravery, as documented in the rhetoric text related to the siege of Dapur.

4

The Manufacture of Protective Gear in New Kingdom Egypt

We bestowed Our favour upon David. (We command): 'O mountains, sing Allah's praises with him'; (and so did We command) the birds. We softened the iron for him, (saying): 'Fashion coats of mail and measure their links with care and act righteously. I am watching over whatever you do.'

(Quran 34.10-11)

4.1 Armories and armourers

As we have seen in previous chapters, in the transition from the Eighteenth Dynasty to the Nineteenth Dynasty, protective clothing progressively loses its exotic and rare status, becoming more widely available among the men serving in the Pharaonic army. In this sense, we might hazard a guess that the manufacture of protective gear began to gain prominence in Egypt starting from the end of the Eighteenth Dynasty. But what do we know about the Egyptian armourers and the places where they used to work?

Again, we must admit that a lack of documentation regarding these aspects is particularly marked. Compared to other manufacturing activities, armour crafting was not popular among the subjects of Egyptian art. Moreover, we do not have any text describing the various stages of the crafting process or a glossary of technical terms relating to defensive equipment.[1] So basically, the few available data turn out to be quite scattered from a chronological and geographical point of view.

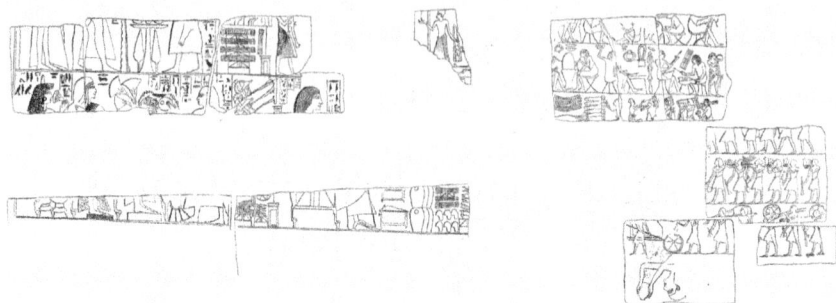

Figure 4.1 Drawing reconstructing a portion of Ky-iry's tomb decoration. Grajetzki, W., *JEOL* 37, 115. (Coutesy of *Journal jaarbericht van het Vooraziatisch – Egyptische Genootschap (Gezelschap) 'Ex Oriente Lux'*.)

There is only very limited iconographic evidence for armour manufacturing during the New Kingdom. The most considerable attestation is represented by the fragmentary decoration of the lost funerary monument of Ky-iry (*PN* I 343, 6), a Nineteenth-Dynasty high official who built his tomb in the New Kingdom necropolis surrounding king Teti pyramid at Saqqara (PM III²/2, 668). At the beginning of the twentieth century, a large number of relief fragments from Ky-iry's tomb chapel were found by the British Egyptologist James Edward Quibell during the excavations of the Coptic monastery of Apa Jeremias at Saqqara, where they had been reused as building material in the foundations of the Southern Church.[2] Starting from these scattered relief fragments, Wolfram Grajetzki has been able to virtually reconstruct the scene depicting the deceased sitting on a chair fitted with leonine legs, while inspecting the outcome of the work of a large group of craftsmen (Figure 4.1).[3] The workers are apparently engaged in a wide range of activities inside *p3 ḫpš* 'The Armoury', probably established in Memphis, where Ky-iry held the ranks of *mr ḥm.t m p3 ḫpš* 'Overseer of the workshop of the Armoury', *mr ḥm.t n Nb-t3.wy* 'Overseer of the royal workshop', *ḥry ḥm.(t) wrrt n p3 ḫpš* 'Head of the workshop of the chariotry of the Armoury', *ḥry ḥmw.w wrrt n* [. . .] 'Head of chariot builders of the [. . .]', *ḥry ḥmw.w wrrt* 'Head of chariot builders', *ḥry ḥmww n p3 ḫpš* 'Head of craftsmen of the Armoury' and *ḥry ḥmw.w / ḥm.t n Nb-t3.wy* 'Head of royal craftsmen/workshop'.[4]

The Papyrus *Anastasi* I (British Museum EA 10247) provides us with a brief description of a productive district surrounding an Egyptian armoury. The relevant passage reads as follows:

Papyrus *Anastasi* I, 26.3–5 (BM EA 10247)

26.3 [hieroglyphs]
26.4 [hieroglyphs]
[hieroglyphs]
26.5 [hieroglyphs]

(26.3) bsy-kwi (26.4) bs.t(i) m ḫn pꜣ ḫpš iw ḳdi tw ḥmw.wt iw ḥmw.w ṯb.w (26.5) mr.wt.n=k nb

'You make your way into the armoury; workshops surround you; craftsmen and leather-workers are all about you. They do all that you wish'[5]

This short excerpt from the Satirical Letter of Papyrus Anastasi I suggests that the activity of the armoury encompasses a wide variety of products. So, it is unsurprising that so many different activities are represented in the five superimposed registers in the right portion of the scene. The upper two registers are crowded with artisans producing precious objects for daily use and funerary equipment. The remaining three registers show rows of men bringing chariots, chariot-related accoutrements and weapons manufactured in the workshop for inspection.[6]

Although no step of the armour-crafting process is clearly shown in Ky-iry's relief, a plethora of helmets and body armour appear among the weapons produced in the workshop (appx. II, III). Behind the deceased, several buildings – perhaps storehouses – contain a multitude of weapons and military paraphernalia. In the lower structure, a scale corslet is arranged on a low table with daggers and a chariot wheel (?). The corslet is shaped like a long shirt with sleeves extended to the elbow and an opening for the head. Although the lower part of the corslet is now lost, we can count at least 125 rectangular scales (closely resembling lamellae) laced in horizontal rows.

As for the helmets, they are depicted in four different contexts: a group of helmets is stored inside the upper building behind Ky-iry; a second group is arranged before Ky-iry, along with shields and bundles of spears (or arrows?), and a third group can be spotted among the weapons the servants are carrying to Ky-iry. The helmets belonging to the fourth and last group deserve a more detailed discussion. On the surface of a non-joining relief fragment, an official of the Armoury is shown weighing a pile of round-topped helmets with the help of a balance. In Egyptian private tombs, weighing scenes usually relate to metallurgical and goldsmithing activities; metal had indeed to be weighted several times during the various stages of production to check the quality of the raw material and to prevent embezzlement.[7] We can therefore assume that the weighing scene depicted here represents the 'quality inspection' of helmets made of metal. We can also suppose that the missing part of the relief showed a scribe writing down the result next to the balance. Although it is not possible to establish a relation between the word *dbn* 'helmet' (Wb V, 438, 1),[8] and the world *dbn* '*deben*, a unit of mass' (Wb V, 438, 2), could a connection exist between the two words sharing the same triliteral root and the weighing scene in Ky-iry's tomb?

Unfortunately for now, the Ky-iry's relief has no analogues among the relatively few known representations of weapon workshops from the Memphite area.[9] Only an unpublished fragment of a limestone stele, kept in the Museum of Fine Arts of Boston (MFA Accession n. 1982.201), could provide a good parallel to the unusual iconography of Ky-iry's relief (Figure 4.2). Since the fragment was purchased on the antiquities market in 1982 by the MFA, its provenance has continued to be unknown. The name of the dedicant is equally mysterious. Investigating this object in more detail, it is not difficult to understand that it was originally the lower right corner of a rectangular round-topped stone slab. The decorative programme of what remains of the stele consists of two registers. In the top register, only the lower extremities of a male and a female figure are still visible. The bottom register, in addition to two fragmentary columns of hieroglyphic text, preserves a section of a larger scene, which shows many similarities with the Ky-iry's relief mentioned above. A significant number of weapons and military equipment are methodically arranged as follows: a group of four khopesh swords, a war chariot, a scale armour (appx. III), a whip, a bundle of arrows (or javelins?) and a group of seven (composite?) bows placed on two small tables. Moreover, on the left, a short-haired man wearing an ankle-length kilt is using a tray for carrying a straight-blade sword. It seems to witness the activities carried out in a military storehouse.

Figure 4.2 Fragment of a Nineteenth Dynasty stele kept in the Museum of Fine Arts, Boston (MFA Accession n 1982.201). Drawing by A. M. Pollastrini.

This wide range of weapons allows us to make further considerations about the place of origin and the dating of the stele. The martial theme of the stone fragment could suggest a connection with the above-mentioned 'Armoury' of Memphis and provide evidence for a possible provenance from the New Kingdom necropolis of Saqqara. As in the case of Ky-iry's relief, a great majority of the weapons and military equipment depicted here, except for the khopesh and the straight-blade sword, relate to chariotry. That is not surprising, in the light of the possible presence in Memphis of a garrison including chariots and horses during the New Kingdom.[10] The war chariot on the relief features six-spoked wheels and a bowcase attached at the right side of the box. Both these features can be used as dating criteria, offering evidence for an attribution to the Nineteenth Dynasty.[11]

A scale corslet is placed under the chariot pole, perhaps to underline the close connection between the two most valuable articles of the Late Bronze Age chariot warrior equipment. The way the corslet has been carved here closely recalls that of Ky-iry's tomb: both are laid out with the opening for the head facing right. Could they have a common iconographical root?

The hierarchical structure and the organization of the craftsmen employed in the manufacture and maintenance of helmets and body armour are not well understood. Only one title relating to armour crafting is known, namely *irw tryn* ⟨hieroglyphs⟩ 'armour maker/armourer' (Figure 4.3). The 'List of People' section of the *Onomasticon of Amenemope*, composed at the end of the Twentieth Dynasty, provides an overview of the personnel employed in weapon manufacturing during the Late Ramesside Period. In this 'classificatory text of "that which exists in the word"', as incisively defined by Ole Herslund,[12] the various workers are probably prioritized according to a hierarchy of importance. *'Irw tryn*[13] is preceded by *tbw nsw* ⟨hieroglyphs⟩ 'King's sandal maker/leatherworker' and followed by *ḥmw mrkbt* ⟨hieroglyphs⟩ 'chariot maker', *ḥmw ꜥḥꜣ.w* ⟨hieroglyphs⟩ 'weapon/arrow maker' and *ir.ty pḏt.w* ⟨hieroglyphs⟩ 'bow maker'.[14] We know from a New Kingdom stele preserved in the University of Pennsylvania Museum of Archaeology and Anthropology (E10996) that a man called Nefer-renpet (*PN* I, 197, 18) held the rank of *irw tryn* in the Nubian fortress of Buhen.[15] In the top register of the stele, the dedicant is shown presenting offerings to the falcon-headed god Horus, lord of Buhen (Figure 4.4), but, except for the title *irw tryn* inscribed over his head, no distinct iconographic element alludes to the armour crafting.

Figure 4.3 'Armour maker/armourer', Papyrus *Golenischeff* (Museum Pushkin 169). Drawing by A. M. Pollastrini.

Figure 4.4 Stele of Nefer-renpet from Buhen (Penn Museum E10996). Drawing by A. M. Pollastrini.

As mentioned above, texts and images suggest that several workshops specializing in the production of all kinds of offensive and defensive weapons – and probably connected to the central administration – were active in the country during the New Kingdom. But is there any archaeological evidence that supports the iconographic and textual data concerning body armour manufacturing?

Possible evidence of armour production in the late Eighteenth Dynasty has been found in the sites of Malqata and Lisht. Malqata is the name applied to the site located at the southern end of the Theban West Bank, where Amenhotep III (c. 1390–1353 BCE) built his monumental residence, 'House of Rejoicing', to celebrate the festivities marking his royal jubilees (PM I/2, 778–781). The compound included no fewer than four palaces, a temple dedicated to Amun, a vast artificial lake, storerooms and servants' quarters. During the excavation undertaken at the beginning of the twentieth century by the Egyptian Expedition of the Metropolitan Museum of Art of New York, the largest group of copper alloy armour scales ever found in Egypt was brought to light.[16] This extraordinary trove, combined with evidence of high-temperature industrial processes (metal, glass and faience working) distributed within and outside the compound enclosure wall,[17] leads to the hypothesis that the Theban Palace of Amenhotep III could encompass one or more manufacturing areas devoted to supplying the demand of the royal court for luxury goods, scale armour included. Unfortunately, for now, the insufficient level of documentation provided by the excavators does not allow us to confirm this conjecture.

The set of armour scales from Malqata (Figure 4.5, 4.6 and 4.7),[18] now kept in the Metropolitan Museum of Art of New York (MMA 11.215.452 a–j), has a uniform appearance.[19] The plates measure approximately 11.5 centimetres in length and have a slightly trapezoidal shape with a slightly rounded upper edge and a roughly right-angled point at the lower edge. Moreover, they have central ribs beaten out from the back, except for the scales MMA 11.215.452 *f* and MMA 11.215.452 *h*, the embossed ribs of which are slightly displaced to the right.

It should be noted that the aim of the embossed rib was not to define the visual appearance of an armour scale but to enhance its solidity and resilience. In other words, the primary purpose of the rib was to prevent the surface of the scale from caving in or denting from a strike. This expedient indeed gave the plate more stiffness for its weight because it was effectively thicker.

Each scale is punched with seven holes to accommodate the rawhide or leather string required to attach the rows of overlapping plates to a heavy fabric or leather substrate.[20] The pattern of holes determines the method of lacing the scales into rows. An old black-and-white picture of the set of bronze scales from Malqata[21] shows fragments of linen cloth adhering to the surface of scales MMA 11.215.452 *a* and MMA 11.215.452 *f*, perhaps suggesting the presence of a fabric undergarment or some kind of textile cover.[22] Strangely, there is a discordance between the formerly mentioned black-and-white picture and the HD images of the individual objects currently provided by the MMA. As it appears today, the fragments of linen cloth seem to have disappeared. Indeed, upon closer inspection, it turns out that the scale MMA 11.215.452 *a* has been photographed turned upside down, not allowing us to detect any textile residue, while the scale MMA 11.215.452 *f* now appears to be totally free from fabric fragments.

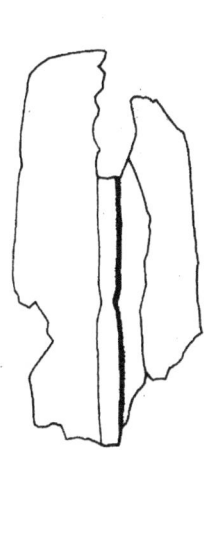

MMA 11.215.452 a

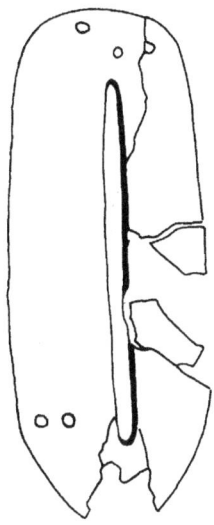

MMA 11.215.452 b

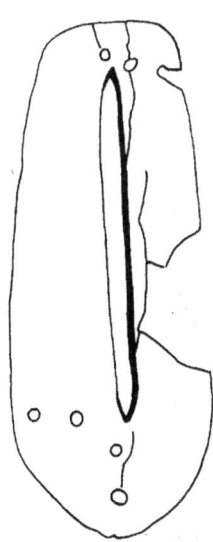

MMA 11.215.452 c

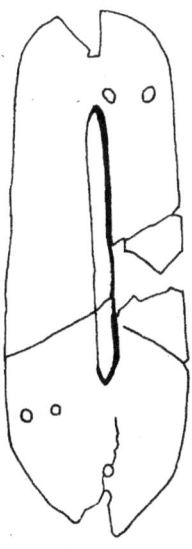

MMA 11.215.452 d

Figure 4.5 Copper alloy armour scales from Malqata (MMA 11.215.452 a–d.) Drawing by A. M. Pollastrini. (Not to scale.)

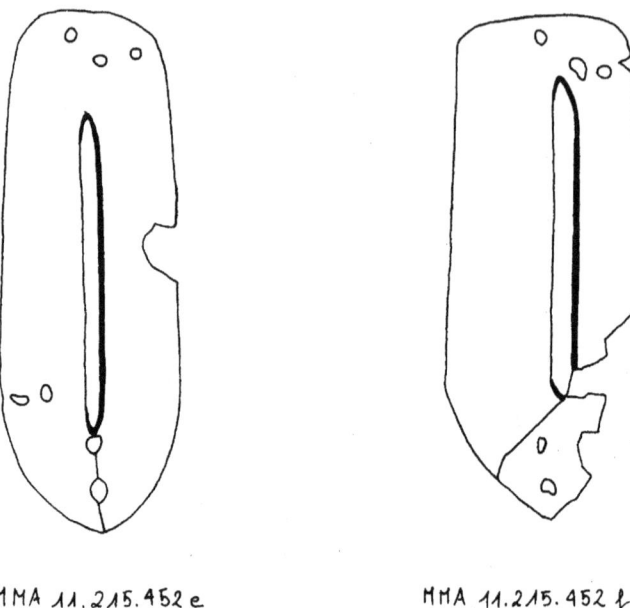
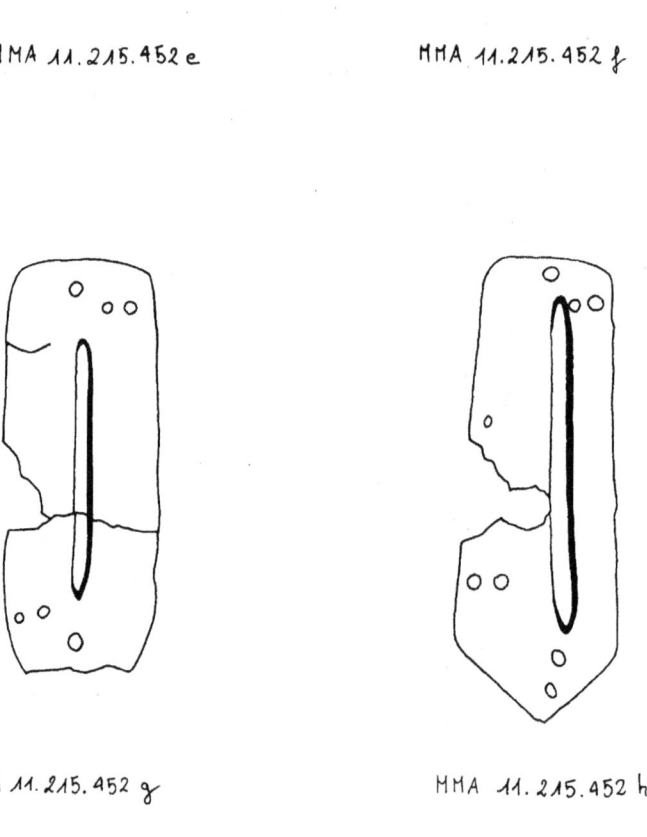

Figure 4.6 Copper alloy armour scales from Malqata (MMA 11.215.452 e–h). Drawing by A. M. Pollastrini. (Not to scale.)

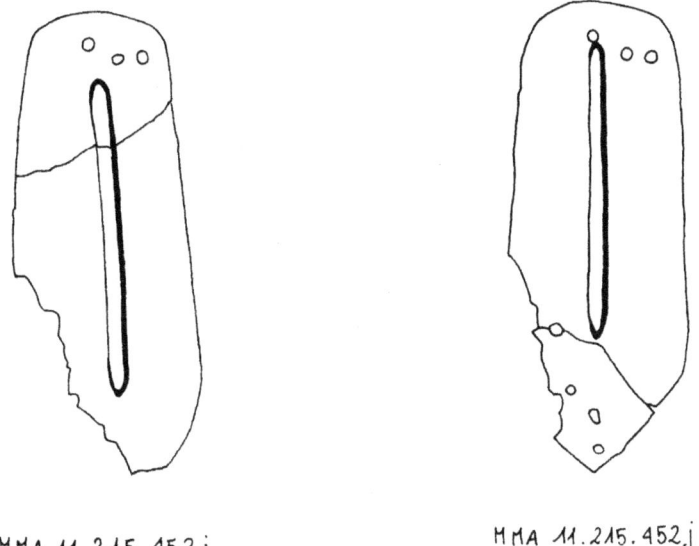

Figure 4.7 Copper alloy armour scales from Malqata (MMA 11.215.452 i–j). Drawing by A. M. Pollastrini. (Not to scale.)

A portion of scale armour found in Room 18 in the House of Prince Šilwa-Tešub at Nuzi could provide a good parallel to the set of armour scales from Malqata.[23] In that area, the archaeological excavations had indeed brought to light a section of thirty-six copper alloy scales fused together in two rows of eighteen, dating back to the period when the city of Nuzi was part of the Hurrian kingdom of Arrapha (fourteenth–fifteenth century BCE).[24] The Type 2 copper alloy scales from Room 18 in Šilwa-Tešub's house (Figure 4.8)[25] and the scales from Malqata (scale MMA 11.215.452 *h* excluded) date back approximately to the same period of time and share some common characteristics: the number and the position of the holes punched on their surfaces, the presence of a central embossed rib and the length of the plates (Type 2 scales from Nuzi = 11.8 centimetres;[26] scales from Malqata = 11.5 centimetres).

As regards the scale MMA 11.215.45 *h*, there was a remarkable similarity between it and a single example of the Type 6 scales from Šilwa-Tešub's house (Figure 4.9).[27] It has been hypothesized that this type of plate was used as a component in the manufacture of scale helmets,[28] or alternatively armoured

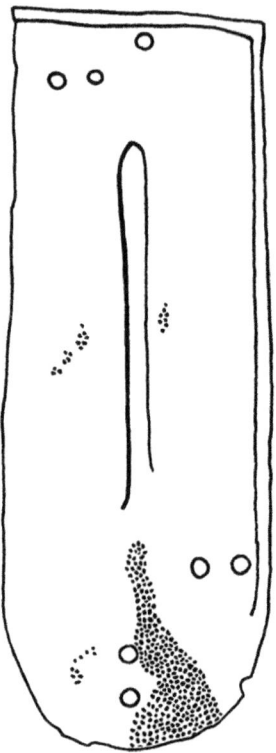

Figure 4.8 Type 2 copper alloy scale from Šilwa-Tešub's house, Nuzi. Drawing by A. M. Pollastrini. (Not to scale.)

skirts.[29] Type 1 armour scales (Figure 4.10)[30] found in the city of Kamidi (present-day Kāmid el-Lōz, Beqaa Valley, Lebanon) also share some similarities with the scale MMA 11.215.452 h. According to Walter Ventzke, Type 1 scales from Kāmid el-Lōz were presumably employed in the production of corslets that reached the wearer's knees.[31]

Turning our attention to the archaeological site of Lisht, we note that possible evidence of armour production is even more tenuous compared to that relating to Malqata.

During the 1933–4 MMA excavations, a reed basket containing a hoard consisting of no fewer than seventy pieces of copper and bronze – broken tools and artefacts, shards and scraps – was accidentally found near the north side of the pyramid of Senusret I (c. 1920–1875 BCE) under a pile of debris.[32] The hoard was wrapped in a piece of cloth sealed with a clay seal bearing the

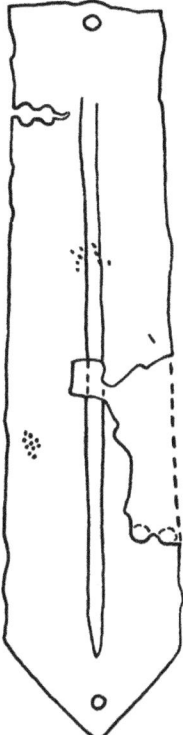

Figure 4.9 Type 6 copper alloy scale from Šilwa-Tešub's house, Nuzi. Drawing by A. M. Pollastrini. (Not to scale.)

praenomen (throne name) of Tutankhamun.[33] The wide range of items found in the basket led scholars to hypothesize that they had been collected by a coppersmith or metal-scrap dealer in order to be reused and recycled due to the high value of the metal. Moreover, W. C. Hayes advanced the hypothesis that the bundle was part of the metalsmith's kit. Although the presence of the royal seal remains basically unexplained, it gives evidence to the fact that the hoard had not been hidden before the end of the Eighteenth Dynasty and was subsequently buried by the debris from the outer casing of the pyramid of Senusret I (PM IV, 81–83) that collapsed.

Remarkably, two copper alloy armour scales (Figure 4.11) are among the many items gathered by the unknown coppersmith.[34] They are now preserved in the Metropolitan Museum of Arts of New York.[35]

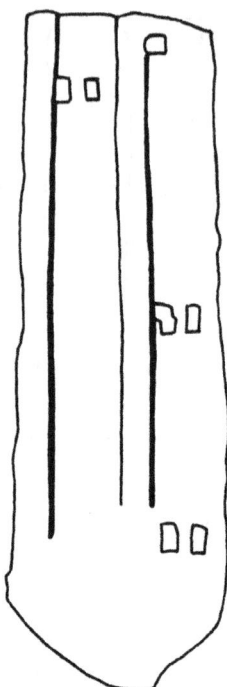

Figure 4.10 Type 1 copper alloy scale from Kamidi (Kāmid el-Lōz). Drawing by A. M. Pollastrini. (Not to scale.)

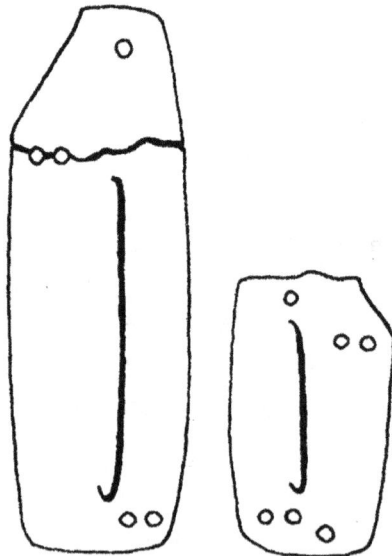

Figure 4.11 Copper alloy armour scales from Lisht (MMA 34.1.72, MMA 34.1.73). Drawing by A. M. Pollastrini. (Not to scale.)

As can be seen from the pictures, scale MMA 34.1.72 is currently broken into two pieces, which measure 8 x 2.5 centimetres and 2 x 2.5 centimetres, respectively. This plate is nearly rectangular with an inward rounded upper left corner. It features an embossed rib slightly displaced to the right. Its surface is punched with five holes distributed as follows: one perforation close to the upper edge, two perforations arranged near the left edge, and two perforations lined up close to the lower right corner.

Scale MMA 34.1.73 is smaller than the previous one (5 centimetres in length; 2.6 centimetres in width). This plate has a slightly trapezoidal shape with an acutely shaped upper edge. It features an embossed rib significantly displaced to the left and shows an array of perforations used for attaching by lacing.

The most conspicuous evidence of armour crafting has been brought to light in the site of Pi-Ramesses (present-day Qantir, Egypt), the Ramesside capital of Egypt from roughly 1300 to 1100 BCE, situated on the Pelusiac branch of the Nile in the eastern Delta. The proximity of the Western Asiatic *theatrum belli* has meant that the city became an essential military base during the Nineteenth and Twentieth Dynasties.[36] The hieratic papyrus *Anastasi* III,[37] dating back to the third regnal year of Merenptah, underlines the strategic importance of Pi-Ramesses, as follows:

Papyrus *Anastasi* III, 7, 5–6 (British Museum EA 10246,1)
'The marshalling place of thy chariotry, the mustering place of thy army, the mooring place of thy ship's troops.'
[Translation according to R. A. Caminos, *Late-Egyptian Miscellanies*, London (1954), 101].

The excavations carried out in area Q I from 1980 to 1987 by the Roemer- und Pelizaeus-Museum in Hildesheim discovered traces of a productive district related to a vast palatial complex. Area Q I is situated south of the modern village of Qantir, not far away (about 250 metres) from area Q IV, where evidence of an extensive horse stable probably relating to the Pharaoh's chariot garrison was unearthed in the 1990s.[38] Four stratigraphical distinct layers have been unearthed in the area Q I:

- D/1: early Eighteenth Dynasty layer;
- B/3: late Eighteenth and/or early Nineteenth Dynasty layer;

- B/2: Nineteenth–Twentieth Dynasties layer;
- B/1: post Ramesside layer.

Stratum B/3, possibly dating back to the beginning of the reign of Ramesses II (or even to the end of the reign of Seti I), contains a production complex comprising two distinct areas: a vast high-temperature metalworking installation in the north and multifunctional workshops where metal, stone, leather, bone and wood were processed, in the south.[39] Following a period of disuse, during the reign of Ramesses II or his successor Merenptah, the metalworking installation was levelled and turned into a court surrounded by octagonal stone pillars which possibly enabled the charioteers stationed in Pi-Ramesses to train and exercise (Stratum B/2). The attached workshops continued to function, supplying everything needed to equip a chariot unit.[40] Stratum B/2 has indeed returned a large number of offensive weapons (daggers, arrowheads and spear points), parts of body armour, components of chariots and horse harnesses, and the craft tools used by artisans employed in the workshops. Moreover, we should also mention the discovery of two fragmentary limestone slabs used as shield moulds for producing metal fittings to be applied on the outer surface of trapezoidal- or 'figure-of-eight'-shaped shields.[41] Considering the Ramesside corpus of battle reliefs, these types of protection resemble those often associated with the Hittites and their Anatolian allies.[42]

A set of ten scales made of different materials (Figure 4.12) has been found in stratum B/2. In addition to four copper alloy scales (FZN 87/0223, 86/0709.0717.0765), archaeologists have unearthed three scales made of bone (FZN 87/0557b, 1-2.1041), two fragments of scale made respectively of yellow (FNZ 86/0661) and red (FNZ 87/0814) glazed ceramic and a plate made of wild boar ivory (FNZ 82/0489).

It must be said that scales made of materials other than copper alloy are unique in the context of the pharaonic hoplology and probably represent attempts to find lighter and cheaper materials to be used in the crafting of armour.[43]

Among the scales found in area Q I, those made in yellow and red faience (FNZ 86/0661, 87/0814) are the only ones that have been published.[44] They have raised ribs and perforations for attaching them to a heavy fabric or leather substrate, looking similar to the metallic scales. Besides, the

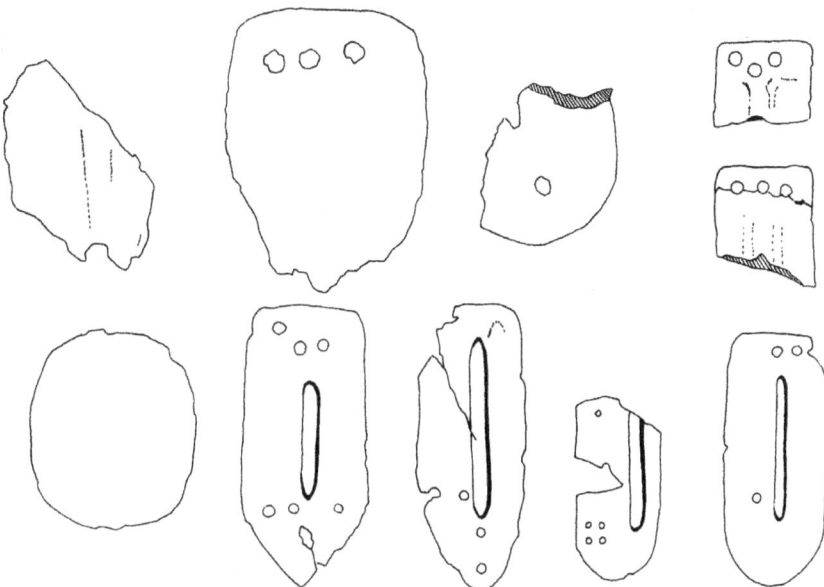

Figure 4.12 Set of scales made of different materials (copper alloy, bone, glazed ceramic and boar tusk ivory) from Qantir, area Q I, stratum B/2. Drawing by A. M. Pollastrini. (Not to scale.)

coloured glaze, which recovers their body, could imitate the lustre of gold and silver.[45]

As regards the little fragment of boar-tusk ivory (FNZ 82/0489) found in the stratum B/2, this has been interpreted with a degree of confidence as part of a boar-tusk helmet, a type of military headwear extensively used in the Mycenean world from the end of the seventeenth to the end of the tenth century BCE. A large body of archaeological and iconographic evidence has enabled scholars to understand the structure of the boar-tusk helmet: it was made of slivers cut from the tusks of a wild boar and attached in rows to a base of some perishable material.[46] Although it would have been hard to assemble, the boar-tusk helmet epitomized the bravery and the hunting skills of the bearer, given that the tusks of at least forty wild boars would have been needed for the crafting of each helmet. The *Iliad* also preserves the memory of this type of military headdress through the description of the helmet given to Odysseus by the Cretan hero Meriones, who had inherited it from his father, Molus:

The Iliad, X, 260-265

> Μηριόνης δ' Ὀδυσῆϊ δίδου βιὸν ἠδὲ φαρέτρην
> καὶ ξίφος, ἀμφὶ δέ οἱ κυνέην κεφαλῆφιν ἔθηκε
> ῥινοῦ ποιητήν· πολέσιν δ' ἔντοσθεν ἱμᾶσιν
> ἐντέτατο στερεῶς· ἔκτοσθε δὲ λευκοὶ ὀδόντες
> ἀργιόδοντος ὑὸς θαμέες ἔχον ἔνθα καὶ ἔνθα
> εὖ καὶ ἐπισταμένως· μέσσῃ δ' ἐνὶ πῖλος ἀρήρει.

And Meriones gave to Odysseus a bow and a quiver and a sword, and about his head he set a helm wrought of hide, and with many a tight-stretched thong was it made stiff within, while without the white teeth of a boar of gleaming tusks were set thick on this side and that, well and cunningly, and within was fixed a lining of felt.
[Translation according to Homer, *The Iliad with an English translation by A. T. Murray*, Vol. 1, Cambridge, MA, London (1924), 454–5].

Scholars often point out that the boar-tusk ivory plate and the above-mentioned shield moulds suggest that foreign artisans worked alongside indigenous craftsmen in the Pi-Ramesses workshops. They possibly supplied non-Egyptian military equipment to the foreign contingents quartered in the Ramesside capital.[47] This small piece of ivory, however, is not the only proof that Aegean warriors – or other 'northern' mercenaries – were hired in the Pharaonic armies during the New Kingdom. The famous illustrated papyrus British Museum EA74100 (Figure 4.13, appx. I),[48] found in the so-called 'House of the King's Statue' at Akhetaten (modern Tell el-Amarna, Egypt),[49] shows a group of Egyptian and Aegean infantrymen fighting against Libyan warriors. According to R. Parkinson and L. Schofield, the parallels in style between the papyrus and the casket n° 21 ('Painted box')[50] from the Theban tomb of Tutankhamun (KV 62) suggest a date towards the end of the Eighteenth Dynasty.[51]

Among the soldiers depicted on the papyrus, the Aegean warriors have been identified through their unusual equipment, which is not present elsewhere in the Egyptian iconographical corpus. While all the figures wear typical Egyptian white kilts, only a smaller number are equipped with short-cropped ox-hide corslets and pale yellow headdresses with vertical red demarcation lines, almost certainly representing Mycenean boar-tusk helmets.

Figure 4.13 Section of the pictorial papyrus British Museum EA 74100 showing Egyptian and Aegean infantrymen. Drawing by A. M. Pollastrini.

4.2 Possible Egyptian copper alloy armour scales discovered outside Egypt

A small number of copper alloy armour scales found beyond the borders of Egypt proper must be added to the total number of findings considered so far. As will be seen in more detail below, there are reasonable grounds for suspecting that they were produced in Egypt or territories under Egyptian control during the New Kingdom.

4.2.1 Beth Shan

As is well known, the Canaanite town of Beth Shan (modern Tell el-Hosn, Israel) served as an Egyptian administrative base and garrison during most of the New Kingdom, from at least the Amarna Period to the Twentieth Dynasty.[52]

Various copper alloy armour scales were discovered in the University of Pennsylvania excavations at Beth Shan (1921–33) in Level VIII–VII contexts, dating back to the Late Bronze Age and related to the long period during which Egyptians were dominant:

- small scale 25.11.501, provided with a rivet (Figure 4.15) was found in Locus 1068 in Level VIII (*c.* 1430–1320 BCE);[53]
- fragmentary scale 27.11.381 (fig 4.15) and scale 27.11.387 (Figure 4.14) were found in Locus 1315 in Level VIII (*c.* 1430–1320 BCE);[54]
- fragmentary scale 25.11.318 (Figure 4.14) was found in Locus 1085 in Level VII (*c.* 1430–1320 BCE);[55]

Added to this is a very fragmentary armour scale found in Locus 1260 in Level VII (*c.* 1430–1320 BCE). Since that scale was later discarded, we have neither information on its appearance nor images depicting it.

More recently, the excavations conducted by the Hebrew University of Jerusalem (1989–96) have unearthed in area Q on the summit of Tel Beth Shan (Stratum Q-2/Level Late VII; Locus 889018) another bronze armour

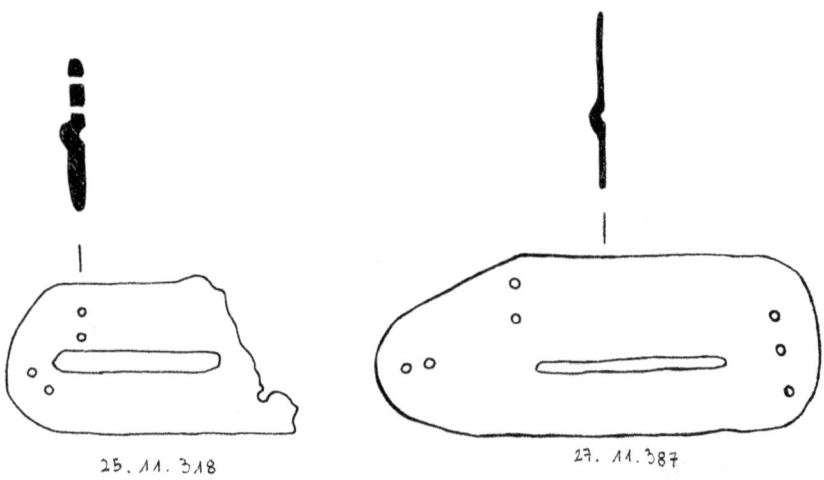

Figure 4.14 Copper alloy armour scales 25.11.318 and 27.11.387 from Beth Shean. Drawing by A. M. Pollastrini. (Not to scale.)

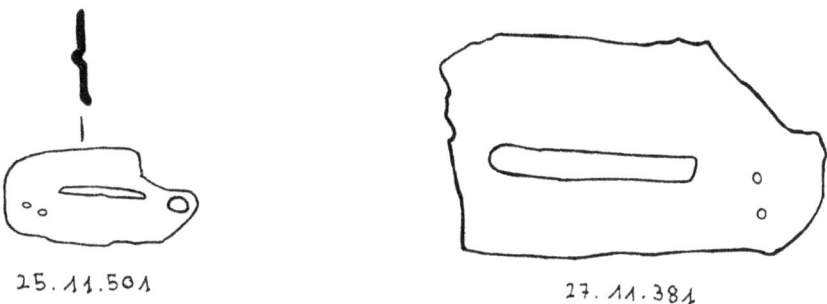

Figure 4.15 Copper alloy armour scales 25.11.501 and 27.11.381 from Beth Shean. Drawing by A. M. Pollastrini. (Not to scale.)

scale (5.9 centimetres in length; 2.5 centimetres in width). This well-preserved plate (Figure 4.16) has a slightly trapezoidal shape with a roughly right-angled point at the lower edge. The embossed rib, which usually occupies the central position of the plate surface, is displaced to the right in this instance. The scale has three perforations arranged in a line along the upper edge and four perforations arranged in a column along the left side.[56]

4.2.2 Kanakia

In 2002, a hoard of bronze objects was found by the Department of History and Archaeology of the University of Ioannina during excavations on the Mycenean acropolis at Kanakia, on the southwest coast of Salamis Island.[57] The hoard had been hidden inside the Buiding IΔ, part of a large complex of structures raised during the first half of the thirteenth century BCE in the eastern part of the Mycenean acropolis and subsequently abandoned around 1200 BCE. Among the bronze objects found in the excavated building was a plate marked with the royal cartouche of Ramesses II, relating to a scale body armour of Near Eastern – Western Asiatic origin.

It should be noted that the findings of Late Bronze Age loose scales are actually very few in number in the Greek mainland. Aside from Kanakia, single examples of bronze plates are known from Mycenae[58] and Tyrins,[59] both datable to the Late Helladic (LH) IIIC period. This limited evidence implies that the scale corselet was never a common article of protective clothing among the Mycenaean warriors.[60]

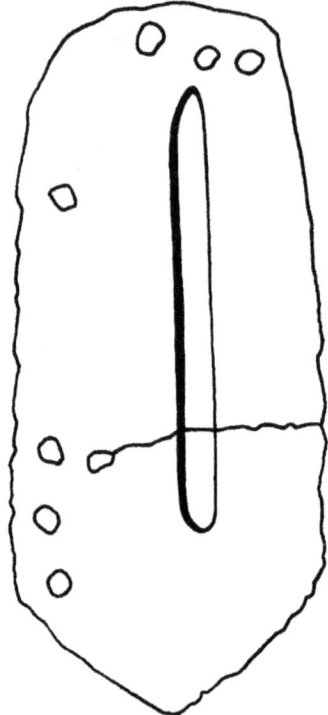

Figure 4.16 Copper alloy armour scale from Beth Shean (Stratum Q-2/Level Late VII; Locus 889018). Drawing by A. M. Pollastrini. (Not to scale.)

The Kanakia scale, 8.5 cm in length, has a rectangular shape with a squared upper edge and a semi-circular lower edge (Figure 4.17). It features an embossed central rib that divides its surface lengthwise into two halves. Seven holes have been punched into the bronze plate: a group of three holes arranged in a row close to the upper edge and a group of four holes displayed in a square-shaped arrangement near the semi-circular lower edge.

But what makes the scale from Kanakia unique is the presence of the cartouche of Ramesses II embossed on the left side of the scale. The cartouche is crowned by a sun disk between a pair of *Atef* feathers and surmounts in turn the sign 𓋞 *nwb* 'gold' (*Gardiner's sign list* S12). According to Catherine Spieser, it is common for Ramesside Period cartouches to display those ornamental features; the sun disk flanked by two ostrich feathers is symbolic of the divine nature of the pharaoh, whereas the hieroglyphic sign for gold perhaps alludes to the supernatural capability of the royal name of shining like

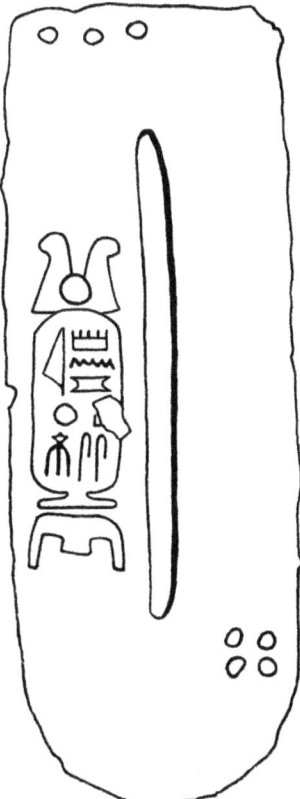

Figure 4.17 Copper alloy armour scale bearing the cartouche of Ramesses II from Kanakia, Salamis Island. Drawing by A. M. Pollastrini. (Not to scale.)

the sun disk.[61] As for the *nomen* (birth name) encircled in the cartouche, it reads R^c-*ms-s(w) mry-'Imn* 'Ra is the one who bore him, beloved of Amon'.[62] The spelling of the birth name of Ramesses II proves to be a useful dating criterion, suggesting that the cartouche was probably inscribed on the bronze scale during the first two decades of the reign of Ramesses II, before the conclusive adoption of the spelling R^c-*ms-sw*.[63]

How did the scale arrive on Salamis island and why? Although many scholars argue that the bronze plate found at Kanakia might be evidence of the employ of Salaminian mercenaries in the Ramesside army or, alternatively, of the Aegean predatory activity perpetrated against Egyptian targets,[64] nothing certain can actually be said about the dynamics through which it reached the island.

4.3 The utilization of leather and rawhide in armour crafting

As we will see later,[65] the hieroglyphic sign ⌒ and its variant form ⌒ (*Gardiner sign-list* F27 and F28), representing a bovine hide, often determine the Egyptian terms relating to defensive equipment. All this 'indirect' evidence hints that the use of animal skin as a component in the manufacture of helmets and armour was a more widespread phenomenon than the current paucity of material remains indicates. It is indeed self-evident that the deterioration of organic materials through the ages has led to the loss of most of the Late Bronze Age protective clothing made of leather or rawhide. With this in mind, the discovery of a cuirass made of rows of overlapping rawhide scales attached to a leather/linen substrate (JdE 62628) in the tomb of Tutankhamun constitutes a one-of-a-kind occurrence. The only existing photograph (p1304) showing this armour in a complete state inside a wooden box (box n. 587) was taken in the immediate aftermath of the discovery of Tutankhamun's tomb by the archaeological photographer Harry Burton. Howard Carter himself described the object as follows:

> Another form of defensive armour was a crumpled-up leather cuirass that had been thrown into a box. This was made up of scales of thick tinted leather worked on to a linen basis, or lining, in the form of a close-fitting bodice without sleeves. It was unfortunately too far decayed for preservation.[66]

Among the hundreds of papers concerning the discovery of the tomb of Tutankhamun, the Griffith Institute Archive holds three object cards written by Howard Carter, which provide us with further information about the Pharaoh's armour:[67]

- Card No. 587a – 1 is the description of the above-mentioned Harry Burton's photograph (p1304).
- Card No. 587a – 2 is a scheme on grid paper showing the scale lacing technique adopted for making the corslet. According to Carter's drawings and captions, we can deduce that the red leather scales have a rectangular shape with a roughly right-angled lower and range from 2.3 to 5.3 cm in length. Besides the single hole along the upper edge designed for fastening the rows of overlapping scales to the backing (one layer of thin leather and

six layers of linen), the scales are provided with a pair of narrow lacing perforations on the left side close to the upper edge. Moreover, the scales feature a vertical running rib slightly displaced to the right.
- Card No. 587a – 3 gives us detailed information on the conservative treatment to which the corslet has been subjected after its discovery:

Treatment for Cuirass – leather 587, A.
Amyl Acetate [and] Acetone mix equal quantities (vol [ume])
Add celluloid to give 2½% soln.
To this add pure castor oil & shake up. Until no further solution
--
Wash finely with celluloid solution if necessary.

Unfortunately, the efforts to consolidate the corslet proved to be inadequate. After being taken to the Egyptian Museum in Cairo together with the majority of the artefacts found in Tutankhamun's tomb, the cuirass was affected by a process of decay, which compromised it irreparably. Currently, the Egyptian

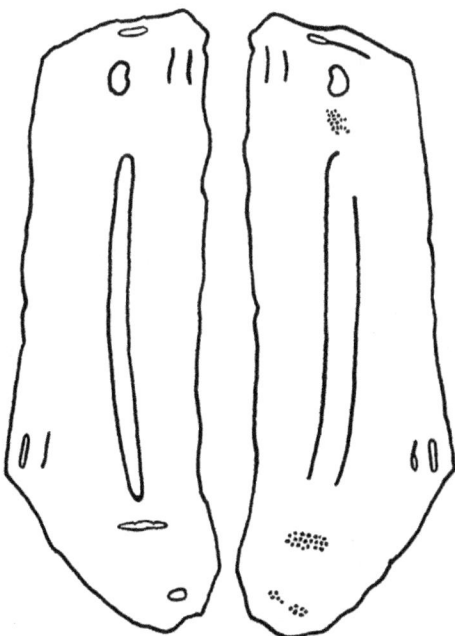

Figure 4.18 Rawhide armour scale from Tell el-Amarna. Drawing by A. M. Pollastrini. (Not to scale.)

Museum preserves only two sections of the armour in which the scales are still laced together and over 200 loose scales. The actual remains have recently been subjected to an interdisciplinary survey which has provided new insights into the variety of types of scales, the lacing techniques, the manufacturing methods and the original appearance of the cuirass belonging to Tutankhamun.[68]

In addition to this exceptionally rare discovery, a single rawhide scale has been found at the site of Tell el-Amarna.[69] This object bears a strong resemblance to some types of scales of Tutankhamun's armour. The scale is roughly rectangular in shape, with a slightly oblique upper edge and a tapered lower edge bent to the left. It measures approximately 6 centimetres in length and 1.7 centimetres in width and has an embossed rib lengthwise down the centre. A single hole, probably designed for securing the scale to the backing, has been punched in the upper half of the scale. There are several additional narrow perforations for lacing: a single horizontal perforation close to the upper edge, a pair of perforations in the upper half of the scale on the left side, a pair of perforations in the lower half of the scale on the right side and a pair of horizontal perforations located in the lower half near the tapered edge.

Further evidence of the employ of leather and rawhide in armour making can be deduced from the hieratic papyrus *Mallet* (Louvre E 11006; Anastasi n.° 1050),[70] dating back to the Years 3 and 4 of the reign of Ramesses IV (c. 1156–1150 BCE). This document consists of six columns of text dealing partly with agricultural affairs. The first column lists the prices of the commodities which the low-ranking officer Imenemoiua received from Thutmes, a servant in the Temple of Khonsu at Thebes, and from another servant named Tchary in a period ranging from the Year 31 of Ramesses III until the Year 3 of Ramesses IV. In line 5, mention is made of a piece of skin (in the text ḥn.t 'skin of a goat or sheep?' (Wb III, 367, 13)) made into a cuirass, which has an estimated value of five *deben* of copper (equal to about half a kilogram of metal):[71]

Papyrus *Mallet*, sheet I, I, 5 (Louvre E 11006)[72]

(I,5) ḥn.t b3k.w m tryn | ir bi3 dbn 5

(I, 5) A piece of hide made into a body armour; it is equal to 5 *deben* of copper.

5

Egyptian Terminology Relating to Protective Gear

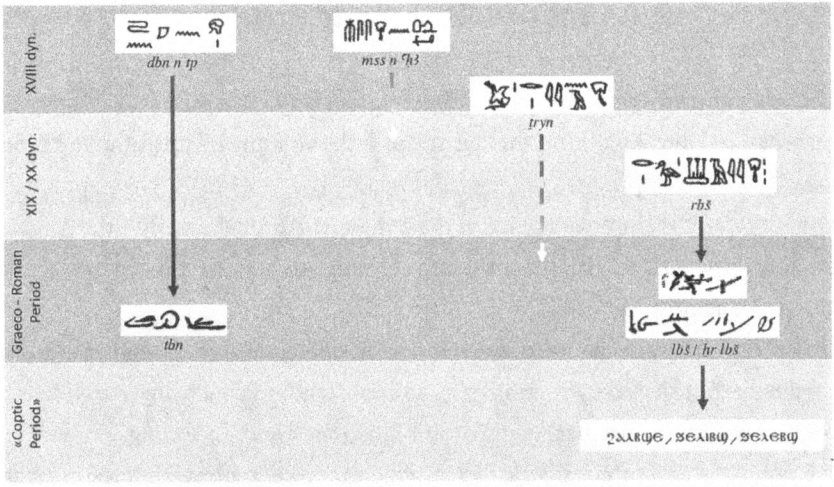

Figure 5.1 Evolution of the Egyptian terms related to the different parts of the defensive panoply (A. M. Pollastrini).

A champion named Goliath, who was from Gath, came out of the Philistine camp. His height was six cubits and a span. He had a bronze helmet on his head and wore a coat of scale armor of bronze weighing five thousand shekels; on his legs he wore bronze greaves, and a bronze javelin was slung on his back.

(1 Samuel 17.4-6)

A side-effect of the spread of combat protective equipment in Egypt during the New Kingdom is the development of a related technical lexicon. Following the transfer of this military technology from Western Asia to their country, the Egyptians needed to introduce new terms into their vocabulary to define them properly.

Although the corpus of attestations is not very large – we have gathered together fewer than forty attestations of words and locutions related to helmets and body armour – we will try to highlight the impact of the introduction of such terminology in the Egyptian lexicon and its evolution through the various stages of the language.

5.1. 𓊵𓂧𓏏𓐍 *dbn n tp / ḏꜣḏꜣ*

The only known Egyptian word utilized to designate 'helmet' is 𓊵𓂧 *dbn*. As regards the New Kingdom, this term can only be found within the locution 𓊵𓂧𓏏𓐍 *dbn n tp/ḏꜣḏꜣ*, '*dbn* of the head' (*Wb* V, 438, 1). The expression *dbn n tp/ḏꜣḏꜣ* is employed twice in the temple of Amon-Ra within the Annals of Thutmose III to list the protective headgears looted during his tenth campaign against Mitanni.[1]

The etymology of the word *dbn* remains an open question. It originates from a triliteral root shared by several verbs and substantives which are related to the concepts of 'being circular', 'going around in circles' and 'encircling'.[2] Moreover, the sign 𓄔 (*Gardiner sign-list* F46), which represent a bovine intestine (the sign 𓄖 *Gardiner sign-list* F48 – variant of 𓄔 – occurs in the two known occurrences of *dbn n tp/ḏꜣḏꜣ*), not only performs an ideographic and phonetic function in 𓄔𓂋 *pḫr*, verb, 'to turn' (*Wb* I, 544–547) and its derivatives, and in 𓂓𓄔 *kꜣb*, verb, 'to double over' (*Wb* V, 8–7) and its derivatives, but it is also employed in 𓅱𓂧𓄔𓂻 *wḏb*, verb, 'to turn' (*Wb* V, 8, 7) and its derivates as determinative. In light of these considerations, therefore, we can assume that the Egyptian language sought to highlight the essential role of the helmet in 'encircling' the wearer's head to protect it from the enemy's blows. Remarkably, the ancient Greek word for 'helmet', περικεφαλαία, emphasizes equally the enveloping 'property' of this kind of protection for the head (περί + κεφαλή, 'around the head'). Περικεφαλαία was absorbed into Coptic (ⲡⲉⲣⲓⲕⲉⲫⲁⲗⲁⲓⲁ) at a later time.

The second part of the locution – *n tp/ḏꜣḏꜣ* 'of the head' – perhaps has the purpose of remarking on the relationship between the object *dbn* and the human head. This detail might have been by no means evident in the period immediately following the introduction of such a new technology.

Both attestations of the term *dbn* in the *Annals* of Thutmose III feature the determinative 𓐍 (Gardiner sign-list N34), representing a metal ingot,[3] which is commonly used in the writing of terms defining objects made of copper or bronze, especially weapons.[4] Moreover, *dbn* is preceded by the substantive 𓎛𓊃 *ḥzmn* 'bronze' (*Wb* III, 163, 14-19) in both instances.

How can this apparent redundancy be explained? An answer to that question could be the propagandistic intent of the *Annals*, which emphasizes rare and precious materials – such as bronze – found in the goods looted by the army of Thutmose III after a military campaign.[5] It should be pointed out, indeed, that in the text of the Annals, the word *ḥzmn* precedes nouns related not only to military equipment but also vessels and other luxury goods.[6]

Moving on to the Late and Graeco-Roman period, it can be noted that *dbn* does not completely disappear but evolves into the Demotic term *tbn* (*DemGl*, 624; CDD T (12.1), 151), following the well-known phenomenon of neutralization between the voiced dental plosive phoneme /d/ and the voiceless dental plosive phoneme /t/.[7] The word *tbn* is attested uniquely in the Demotic fictional narratives composing the epic cycle of Inaros of Athribis:[8] twice in the *The Conflict over the Benefice of Amun* (P. Spiegelberg + P. Ricci, first century BCE)[9] and once in *Prince Petechons and Queen Serpot* (P. Vienna D 6165 + D 6165 A, c. 200 CE).[10] Furthermore, it must also be said that tbn, according to the textual sources at our disposal, is never followed – unlike *dbn* – by the locution *n tp/ḏ₃ḏ₃*, proving that during the late stage of the Egyptian language, the use of such expression is no longer considered necessary to convey the exact meaning of this term.

5.2 𓅓𓋴𓋴𓏤𓐛𓂝𓎛𓄿 *mss n ʿḥ₃*

As in the previous case, the earliest written occurrence of body armour dates back to the Eighteenth Dynasty and, more specifically, to the reign of Thutmose III.[11] The periphrasis 𓅓𓋴𓋴𓏤𓐛𓂝𓎛𓄿 *mss n ʿḥ₃* 'combat smock' (Wb II, 149, 7) was created employing available common words with the purpose of integrating the Egyptian vocabulary with a new concept[12] related to a previously unknown foreign military technology. Another significant circumlocution of this kind is represented by the *hapax legomenon* 〈hieroglyphs〉 ʿhʿ.w ʿḥꜣ.w m ḥry – ib pꜣ Ym, literally 'ships of the warriors from the midst of the sea', employed in the text of rhetorical stele of Ramesses II from Tanis (Cairo Museum Cat. 34510) to describe the unfamiliar warships operated by Sherden warriors and used for make raids.[13]

Only ten occurrences of this circumlocution are currently known:

- eight are in the text of the *Annals* of Thutmose III, distributed as follows:
  - three occurrences in Section I, inscribed on the north wall of the passage around the granite Naos of Philippus Arrhidaeus, Karnak Temple complex PM II², 97, [280];[14]
  - three occurrences in Section V, inscribed on a fragment belonging to the north wall of the Vestibule of Thutmose III between the Sixth Pylon and the Central Court, Karnak Temple complex. The fragment is now kept in Louvre Museum (C 51; N 205; Salt n° 3834), PM II², 89, [240]–[244];[15]
  - two occurrences in Section VI, inscribed on the inner face of the Sixth Pylon PM II², 89, [240]–[244];[16]
- one in the Victory stele of Thutmose III from Gebel Barkal. The stele is now kept in the Museum of Fine Arts of Boston (Accession number 23.733), PM VII, 217, [20];[17]
- one in the stele of Amenophis II, placed in front of the south face of the Eighth Pylon, PM II², 177, (R).[18]

The first element of the circumlocution shares with the substantive 〈hieroglyphs〉 *mss(.t)* 'smock, tunic' (*Wb* II, 149, 8) the same triliteral root. What actually differs between the two substantives is the semagram.[19] The sign 〈sign〉 and its variant form 〈sign〉 (*Gardiner sign-list* F27 and F28), both representing a cowhide, can be found in most of the occurrences listed above, instead of the sign 〈sign〉 (*Gardiner sign-list* V6), which is pertinent to the 'Clothing made up of textile fibres' lexical field. The determinative 〈sign〉 clearly focuses on the use of animal skin in manufacturing body armour, broadly confirming the theory by which leather (or even rawhide) would be the most suitable material for making tunics which do not deform under the weight of the scales sewn on it.[20] Among the occurrences of the word *mss* gathered here, only one of those extrapolated from the *Annals* of Thutmose III (Urk IV, 732, 1) features the determinative 𓈒 (*Gardiner sign-list* N34), representing a metal ingot.

5.3 𓌉𓏤𓈖𓈙𓂋 *ṯryn*

Following the development and diffusion of Late Egyptian during the end of the Eighteenth Dynasty and the Ramesside Period,[21] an increasing number of foreign words were introduced into the Egyptian language.[22] Such is the case of 𓌉𓏤𓈖𓈙𓂋 *ṯryn* 'body armour' (*Wb* V, 386, 6–10), a word probably derived from the Hurrian term *šarianni/šariyanni* 'body armour'.[23] *Šarianni/šariyanni* was incorporated as a loanword into Hittite vocabulary and also in some other languages belonging to the Semitic family: *šariyanni*- 'scale armour' (Hittite),[24] *siriam* 'cuirass' (Akkadian),[25] *zariam* 'cuirass' (Nuzi),[26] *ṯryn* 'cuirass' (Ugaritic),[27] *š/siryō/ān* 'cuirass' (Hebrew).[28] The attempt to relate this term to the Arabic word *jirān* 'frontal part of the camel neck' does not seem to be supported by any substantial evidence.[29]

Are there other elements to better understand the dynamic processes that led to the introduction of the word *ṯryn* in the Egyptian vocabulary? In answering this question, we should give prominence to a specific text belonging to the diplomatic archive found at the archaeological site of Tell el-Amarna (Akhetaton, 'the horizon of the Aten', the capital city of the late Eighteenth Dynasty). The clay tablet EA 22 (Staatliche Museen zu Berlin, Vorderasiatisches Museum Ident. Nr. VAT 00395) was sent to the Egyptian king Amenhotep III by the king of Mitanni Tushratta on the occasion of the marriage of his daughter Tadu-hepa to the Pharaoh himself, in order to give an accurate list of the wedding gifts he presented to the couple. EA 22, like most tablets from Tell el-Amarna, is written in Akkadian, the diplomatic *lingua franca* adopted in the official correspondence by Egypt and many other States in the ancient Near East, roughly between the fourteenth and the thirteenth centuries BCE.[30] The following excerpt (EA 22, III, 37–41) is of particular interest:

37. 1 cuirass set, of bronze. 1 helmet, of bronze, [f]or a man
38. 1 cuirass set, of leather. 1 helmet, [of br]onze

39. for the *sarku*-soldiers. 1 cuirass set, of leather
40. for horses, set with ri[ng]s of bronze
41. 2 helmets (chanfron?), of bronze, f[or ho]rses
[Translation according to W. L. Moran, *The Amarna Letters*, Baltimore, London, (1992), 55].

In the excerpt quoted above, the Akkadian word for 'cuirass' appears three times: twice as *sariam* (EA 22, III, 37–38) and one time as *zariam* (EA 22, III, 39). Furthermore, it should be noted that the same term is employed to define not only body armour for men but also the armour designed for horses. Considering this textual source, can we assume that the word *ṯryn* has been borrowed from Hurrian through the medium of the diplomatic Akkadian?

An article by Philippe Collombert and Laurent Coulon[31] has highlighted an occurrence of the word *ṯryn* in a hieratic papyrus fragment (pBN202) preserved in the *Bibliothèque Nationale de France*, which has been recognized as the *incipit* of an epic concerning the fight between the gods and the deified Sea. This mythological tale only survived on the extremely fragmentary papyrus *Amherst IX*,[32] also known as Astarte Papyrus, preserved in the Pierpont Morgan Library of New York. Thanks to an in-depth analysis of the papyrus fragments, the two scholars determined that pBN202 and papyrus *Amherst IX* were originally parts of the same scroll.[33] According to palaeographical and grammatical examinations, the text has been unequivocally dated to the reign of Amenhotep II. Thus, the occurrence of *ṯryn*, taken into account here, must be considered, as far as we know, the earliest recorded use of this particular term in the corpus of Egyptian literature.

The date assigned to the papyrus pBN202 + *Amherst IX* by Philippe Collombert and Laurent Coulon provides evidence of the coexistence of the locution *mss n ꜥḥꜣ* and the word *ṯryn* during the reign of Amenhotep II.

Far from being a minor issue, this point allows us to verify the ongoing process of introducing emerging concepts into Egyptian vocabulary. This process could be roughly divided into two distinct temporal phases: the first consisted of filling the lexical gap by coining a neologism or a circumlocution in an effort to evoke a specific idea; the second consisted of adopting more adequate words from a different language (borrowings) which subsequently existed side-by-side or replaced the previous expressions.

This phenomenon is not isolated. Two terms, for instance, were used to identify the horse:[34] a genuine Egyptian word 𓎛𓏏𓂋𓃻 *ḥtr* (*Wb* III, 199–200) and a foreign loanword 𓋴𓋴𓅓𓏏𓃻 *ssm.t* (*Wb* IV, 276–277).[35] Another peculiar case is represented by the sickle-shaped sword. At an earlier stage, in the Annals of Amenemhat II,[36] the new weapon was merely called 𓌢 *ꜣsḫ* 'sickle' (*Wb* I, 19), using a descriptive term which underlines the morphology of a previously almost unknown object; at a later stage, after it was firmly established in the royal ideology, the sickle-shaped sword changed its name in 𓂝𓄹 'human arm' or 'animal limb' (*Wb* III, 268–270), a word of Egyptian origin.

Getting back to the coexistence of *mss n ꜥḥꜣ* and *ṯryn* during the reign of Amenhotep II, it should be made clear that they were employed in different contexts, perhaps suggesting two different linguistic registers. The circumlocution of *mss n ꜥḥꜣ* was used in an official text engraved on a stone stele erected in the Precinct of Amun-Ra at Karnak. The word *ṯryn*, however, was used in a literary text – a panegyric dedicated to Amenhotep II showing a certain similarity with the genre called *sḏd nḫtw* 'tale about victories' – written with a language which blends Late Egyptian and traditional Middle Egyptian elements.[37]

Nevertheless, we can reasonably assume that *ṯryn* was totally assimilated into Egyptian military terminology and also adopted in the lexicon a phraseology related to the Pharaoh, replacing the previous expression *mss n ꜥḥꜣ*, not later than the Nineteenth Dynasty.[38] Most of the known occurrences of the word *ṯryn* are actually in the historical – narrative texts on the military campaigns of Ramesses II. They are distributed as follows:

- one in the copy of the *Poem* (K₁) of the battle of Kadesh, engraved in Karnak Temple complex (PM II², 58, [174]);[39]
- one in the copy of the *Poem* (L₁)[40] and 1 in the copy of the *Bulletin* (L₁)[41] of the battle of Kadesh, engraved on the outer face of the entrance Pylon, Luxor Temple (PM II², 304–305, [13]–[14]);
- one in the copy of the *Poem* (L₂)[42] and 1 in the copy of the *Bulletin* (L₂)[43] of the battle of Kadesh, engraved on the outer face of the eastern and southern walls of Ramesses II's Court, Luxor Temple (PM II², 335, [216]–[218]);
- four in the rhetoric text (L) related to the scene of the siege of Dapur, engraved on the outer face of the western wall of Ramesses II's Court, Luxor Temple (PM II², 333, [202]–[203]);[44]

- one in the copy of the *Bulletin* (R₁) engraved on the inner face of the 1st Pylon, Ramesseum (PM II², 433, [3]);[45]
- four in the rhetoric text (R) related to the scene of the siege of Dapur, engraved on the entrance wall of the hypostyle hall, Ramesseum (PM II², 438, [17]–[18]);[46]
- one in the copy of the *Bulletin* (I) engraved on the north wall of the Great Hypostyle Hall, Great Temple of Abu Simbel (PM VII, 103–104, [41]–[42]);[47]
- two in the partial copies of the *Poem* (Ch B₁, Ch B₂) written on the recto and verso of the hieratic papyrus *Chester Beatty III* (BM EA10683,2);[48]
- one in the copy of *Poem* (Rf – S) written on the hieratic papyrus *Raifé/Sallier III* (pRaifé = Louvre E4892, pSallier III = BM EA10181).[49]

The remaining occurrences of the word *ṯryn* can be found in some documents (Papyrus *Mallet*, *Onomasticon Golenischeff* and Stele Penn Museum E10996), relating to the manufacture of armour and helmets, which are described in more detail in Chapter '4. The Manufacture of Protective Gear in New Kingdom Egypt'.

To better understand the transition of the loanword *ṯryn* into the Egyptian vocabulary, we must focus on certain phonological and writing issues. First, it is crucial to highlight the well-known correspondence between the Egyptian phoneme /ṯ/ and the Semitic phoneme /s/.[50] In addition to that, if we keep in mind the proximity between the palatal plosive phonemes /ṯ/ (voiceless) and /ḏ/ (voiced),[51] along with the relation linking the Egyptian phoneme /ḏ/ and the Semitic phonemes /š/z/ṣ/, the transition of the word *ṯryn* from the Semitic languages to the Egyptian become more transparent.

So it is not surprising that, in an attempt to record the unfamiliar Semitic phonemes, scribes wrote the word *ṯryn* employing the system known as 'group writing' or 'syllabic writing', in which a new set of sign groups was used to write syllables, replacing the traditional Middle Egyptian orthography.[52] Many scholars have tried to establish whether 'group writing' was used on regular basis to note foreign – especially Semitic – loan words, but the final word on that point is far from being said.[53]

As for the known occurrences of *ṯryn* relating to the monumental engravings as well as the manuscript texts, they show a certain degree of consistency.

Apart from a few slight exceptions, the word is usually written 𓆱𓏤𓂋𓇋𓇋𓈖𓏌𓏐. As might be expected, hieratic texts provide some variants: 𓆱𓏤𓄿𓇋𓇋𓈖𓏌𓏐 in papyri *Chester Beatty III* (BM EA 10683,2) and *Sallier III* (BM EA 10181); 𓂋𓂝𓄿𓇋𓈖𓏐𓏤 in papyrus *Mallet* (Louvre 1050 = E.11006).

William Albright, in his obsolete study on Egyptian syllabic orthography, endeavoured to restore the vocalization of *ṯryn* as follows: *Ti – ir(ra) – ya – na*.[54] However, it should be said that, although outdated, the work of Albright was later mentioned by Timothy Kendall[55] and Tamás Dezsö.[56] Following the scheme proposed by Friedrich Junge we can try to dissect the word *ṯryn* as follows: 𓆱𓏤 *ṯꜣ* > *ṯ*; 𓂋𓂝 *rꜥ* > *r*; 𓇋𓇋 *y*; 𓈖 *nꜣ* > *n* = *ṯryn*.

5.3.1 An armoured goddess?

Discovered by Selim Hassan at Giza,[57] near the place of worship improperly called 'Temple of the Sphinx,'[58] the New Kingdom stele of Pay, measurer of the god Hauron (Cairo Museum J. d. E. 72289) bears a unique representation. In the upper part of the stele are carved three standing deities: a falcon-headed god, a naked god wearing the sidelock of youth and bearing a bow and an arrow, and a goddess with long hair, wearing a long dress of foreign origin. The names of those gods are invoked by the donor in the formula written in the lower field of the stele: 'Pa-Shed', 'Isis, the Great', and 'Horus, the Son of Isis'. If, on one hand, the identity of the male deities is basically confirmed by the epithets written above their heads ('Shed, the Great God, the Lord of Sky, the excellent physician, beloved of Egypt', and 'the Son of Isis, sweet of love'), on the other hand, the interpretation of the female member of the triad, remains controversial, mainly due to her singular aspect and the enigmatic epithet related to her.

Unlike the traditional iconography of Isis,[59] the goddess depicted here is characterized by a hairstyle and a dress usually related to Asiatic women in Egyptian art.[60] At the same time, the epithet written above her head is still not clearly understood. According to Selim Hassan, Rainer Stadelmann and Christiane Zivie – Coche, this *hapax legomenon* should be read as 𓂋?𓏤𓏤𓇋𓇋𓂝?𓈖? or 𓂋𓏤𓏤𓇋𓇋𓂝?𓈖? *Mtryw* 'Meteryu'. Nevertheless, scholars have, unfortunately, not been able to provide a translation of it.[61]

Vladimir Vikentiev has assumed that the goddess on the stele represents a particular manifestation of Isis, bearing the epithet 𓂋𓏤𓏤𓇋𓇋𓂝𓈖. Curiously enough, the scholar translated this epithet as 'Terre des conjurations de repoussage'.[62]

Finally, Ali Radwan explained this unique epithet, not without some difficulty, as a variant spelling of the word 𓋴𓏭𓈇𓈖𓏌𓏲𓂝 *ṯryn*, alluding to the role of the goddess as 'divine cuirass' which keeps the worshipper safe from evil.[63] Bearing in mind the possible Asiatic origin of this particular manifestation of Isis, a question arises: is it possible to recognize a female deity sharing similar characteristics in contemporary foreign pantheons?

In an attempt to answer this question, or at least to get closer to an answer, we should take into consideration Shaushka, the Hurrian equivalent of Akkadian-Babylonian goddess Ishtar.[64] Her cult had an enormous influence in Northern Syria and Northern Mesopotamia during the second millennium BCE. Following the decisive victory of the Hittite king Suppiluliuma I (c. 1344–1322 BCE) over the kingdom of Mitanni, Shaushka assumed increasing importance in Anatolia until she became the personal protector of the king Hattusili III (c. 1267–1237 BCE).[65] Textual and iconographic sources underline her ambiguous sexual nature, which finds an explanation in her double function as goddess of love (including carnal love) and war.[66] In this respect, Hurrian-Hittite texts[67] from the site of the Hittite capital Hattusa (modern Boğazkale, Turkey), relating to a hypostasis of Ishtar/Shawshka worshipped at Šamuha,[68] provide information on the panoply belonging to the goddess. Further to the mace, the axe, the quiver, the bows and arrows, the body armour (*šarianni*) and a helmet (*gurpiši*) are associated with Shawshka (KUB 27.1 II 9; 27.6 I 18).[69]

Could there be a relation between Shawshka of Šamuha and the hypostasis of Isis depicted on Pay's stele?

5.4 𓂋𓃀𓈙 *rbš* / *lbš*,

According to the available textual documentation, the word *ṯryn* seems to have been replaced by the term 𓂋𓃀𓈙 *rbš* 'body armour' (*Wb* II, 414, 6) in the later part of New Kingdom. Once again, we are dealing with a word of foreign origin borrowed by the Egyptian language and integrated into the lexicon of military-technical terms.[70] As in the previous case, the word *rbš*, related to the Semitic root *lbš* (eg Ugaritic *lbš, lpš* 'garment'),[71] has been almost entirely written in syllabic writing, following a phenomenon peculiar to the Late Egyptian: ⌒ı *rꜥ* > *r*; 𓅓 *b₃* > *b*; 𓈖𓈖 *š₃* > *š*.[72] As for the semagram 𓍢 (*Gardiner sign – list* F28), it indicates that *rbš* is part of the semantic category related to the objects made with animal skin. Furthermore, the plural strokes 𓏤 (*Gardiner sign – list* Z3) have been added without regard for the number of the term, in a way that was actually quite common in the Late Egyptian writing system.

The oldest known occurrence of *rbš* can be found in the Papyrus *Koller* (Pap. Berlin 3043), which dates back to the Nineteenth Dynasty.[73] This manuscript comprises four hieratic model letters. The term *rbš* appears in the one known as 'The equipment of a Syrian expedition' which occupies the first column and the beginning of the second one of the *recto* of the papyrus. The text of this short model letter is basically a list[74] of horses, stable attendants, grooms, chariot harnesses and weapons that must be provided for a military expedition to Syria.[75]

The longevity of the word *rbš* is somewhat unique in comparison with the lifespan of the other terms taken into account here. In the Demotic tales of the epic cycle of Inaros of Athribis, where warfare plays a key role, it is not uncommon to come across the word *rbš* evolved into the form *lbš* / *lybš* (*DemGl*, 262; *CDD* L (01.1), 6). Occurrences of this term can be found in the following tales:

- *Inaros and the Griffon* (P. Carlsberg 80, first/second century CE);[76]
- *The Conflict over the Benefice of Amun* (P. Spiegelberg + P. Ricci, first century BCE);[77]
- *The Contest for the Armour of Inaros* (P. Krall, 137/138 CE);[78]
- *Prince Petechons and Queen Serpot* also known as *Egyptians and Amazons* (P. Vienna D 6165 + D 6165 A, c. 200 CE).[79]

The use of *rbš* did not end with the Demotic stage of the Egyptian language. This term survived down to Coptic, through the words (L) ϩⲁⲗⲃϣⲉ, (B) ϩⲉⲗ(ⲗ)ⲓⲃϣ · ϩⲉⲗⲉⲃϣ 'body armour,'[80] which probably originate from the Demotic variants *ẖ-lybš* / *ẖr-lbš* (*DemGl*, 262; *CDD* L (01.1).[81] However, it does seem that the old Semitic loanword coexisted with the new Greek loan word ⲡⲁⲛϩⲟⲡⲗⲓⲁ 'πανοπλία, suit of armour'.[82]

Mention should be made of the verb *rbš*, 'to wear a body armour',[83] belonging to the same morphological family. To date only two occurrences of this verb are known. The older one lies within the text of the Hieratic administrative papyrus Louvre AF 6347, dating back to the end of the New Kingdom. This manuscript is actually an official record concerning the measurement of fields which belonged to the Domain of Amun around Thebes.[84] On the recto of the papyrus (A, 14)[85], the verb *rbš* is part of the theophoric anthroponym 𓀀𓏏𓏤𓂋𓃀𓈙𓏛 (*Ḥr rbš* (?), the writing of which poses problems of interpretation. On the one hand, if we consider the second part of the name as a pseudo-participle of the verb *rbš*, we may translate it as 'Horus is armoured'; on the other hand, if *rbš* is a noun, we may translate it as 'Horus (is my) armour'. Nevertheless, in the latter case, one would have expected the writing *Ḥr m rbš* (=*i*), although it is conceivable that the *m* could be intentionally omitted by the scribe.

The second occurrence of the verb *rbš* is in a Magical Papyrus from the Brooklyn Museum (Accession n 47.218.516). This fragmentary manuscript, dating back to the first millennium BCE after the Saite Period, is inscribed with two hieratic magical spells intended to be used against unspecified dangerous enemies. In addition, two vignettes show the divine entities summoned to seek protection. A passage from the second spell describes the physical features of one of these deities: among other things, the benevolent god is described as 𓂋𓃀𓈙𓂻 *rbš* 'armoured' (Pap. Mag. Brooklyn 5, 2).[86] Here again, from a grammatical point of view, we are probably facing a pseudo-particle of the verb *rbš*. The vignette related to this particular spell depicts a heterogeneous figure which resembles the god Bes, combining animal and symbolic elements with apotropaic function.[87] The body of the deity is covered with a long shirt made from a dotted pelt: is that garment what makes him 'armoured' against the evil?

5.5 The verb 𓏏𓍿𓄿 *ṯȝi*

Once again, we have to admit that the lack of data also affects our knowledge of Egyptian terminology relating to the use of armour. Only the large corpus of Ramesside narrative texts provides us with a few references to suits of armour worn by combatants in real combat action.

As mentioned above in section '3.2 Body armour in Ramesside royal ideology', the inscriptions relating to the battle of Kadesh (*Poem* and *Bulletin*) and the rhetorical text connected to the siege of the Syrian town of Dapur inform us that Ramesses II donned his armour in both confrontations. In both instances, the action of wearing the armour is expressed by the *tertiae infirmae* verb 𓍿𓄿𓇋𓂡 *ṯꜣi*, 'to don, to wear' (*Wb* V, 246, 5). Given the limited textual evidence, can we reasonably assume that the verb *ṯꜣi* was adopted in the military lexicon developed during the New Kingdom?

6

Conclusions

In the beginning, the primary goal of this study was practical: to collect all the pieces of evidence known up to the present concerning the use of corporal defensive armament in Egypt during the New Kingdom to shed light on a topic that has not received much attention. The choice was dictated by the need to highlight the role played by this typology of military equipment in the evolution of warfare that occurred during the second millennium BCE in the eastern Mediterranean basin.

The success of new combat tactics, relating to the use of chariots drawn by two horses as well as the composite bow, following the gradual expansion of the Khurrite-Mitannians in the vast region of Upper Mesopotamia from the Mediterranean to the Zagros Mountains, stimulated the adoption of different types of coverings which allowed warriors to have their hands free to fight and to drive their vehicles.

Within the framework of Egyptological studies, few works have been dedicated to personal protective equipment, in contrast to the chariot and composite bow. With this in mind, I have examined the subject through different approaches – iconographic, archaeological and lexicographic – with the aim of 'crossing' data from independent fields, trying to obtain an overview as close as possible to the 'truth'. Immediately, this perhaps simplistic and distorted view of how to develop research collided with several issues.

First, the scarcity of sources – except iconographic ones – constitutes a fundamental impediment to the correct understanding of the dynamics that characterized the introduction, diffusion and use of the new corporal defensive equipment in New Kingdom Egypt. We have deliberately used the term 'new' here in relation to Egypt because we have found no convincing evidence of the use of armour before the Eighteenth Dynasty. Unfortunately, the scarcity of evidence at our disposal prevents an accurate analysis of the phenomenon.

Secondly, this approach, which was intended to focus on well-defined temporal and geographical limits, paradoxically required considerable work on sources dating back from different eras and concerning territories beyond Egypt's borders. Nevertheless, this is the only way to obtain an exhaustive overview of the evolution of protective clothing at a regional level.

From a chronological perspective, we have devoted part of the work to the prospecting of testimonies of corporal defensive equipment, which embody the antecedents of helmets and corslets of the Late Bronze Age, extending our field of research to the third millennium BCE.

From a geographical perspective, in light of the well-known network of commercial and cultural exchanges characterizing the eastern basin of the Mediterranean during the Late Bronze Age, we have not overlooked the many foreign influences exercised not only in the technological field but also in ideological and lexicographical ones. In this regard, concerning the morphological aspects of body armour, we have taken into account the most significant archaeological evidence, which the sites of the Near East, Anatolia, and continental and insular Greece have delivered, to have the possibility of identifying relevant parallels – essential to the approach of our study – which the exiguity of the Egyptian testimonies denies us.

Moreover, concerning the lexicographic aspects, I have focused on the dynamics which characterized the passage of foreign terms of Semitic origin into the Egyptian military language and the consequent disappearance of the autochthonous periphrases, which were precociously elaborated to integrate into the Egyptian vocabulary several new concepts, related to foreign military technology.

As we noted previously, archaeological and written sources remain scarce, but the help of representations allows us to define the main trends in the equipment of the Pharaonic soldier.

At the beginning of the Eighteenth Dynasty, soldiers inherited traditional Egyptian defensive equipment: the rectangular or trapezoidal shield with a rounded top, typical of the infantry. On the other hand, helmets and corslets, in iconographic sources, are relegated to representations of defeated Asian enemies, perhaps with an ethnic connotation.

From the Thutmoside period, as a result of several Asiatic military expeditions, a certain quantity of weapons, including helmets characterized by a conical or oval-shaped skull and scale corslets, flowed towards Egypt in the form of booty, gifts and tributes, as testified by texts and pictorial representation. Moreover, in military scenes, corslets and helmets – raised from a single piece of metal or made of scales – are often worn by the Western Asiatic warriors defeated and subjugated by the Pharaoh. Not infrequently, in the same scenes, helmets are depicted lying on the ground among the battlefield debris.

It is necessary to wait for the Amarna period and the end of the Eighteenth Dynasty to see a modest diffusion of helmets exclusively among the Egyptian charioteers. Conical skulls, sometimes provided with a plume at the top, characterize these head coverings. On the other hand, scale cuirasses, even the less expensive leather models, must have been reserved for the high classes (the sovereign and the officers of his entourage).

The Ramesside period saw increased production of arms and armour in highly specialized workshops to meet the army's needs. The discovery of archaeological evidence of well-known types of Asian and Aegean defensive equipment in the workshops of the city of Pi-Ramses also raises questions about the impact of foreign labour on the production of military equipment in Egypt.

Judging by iconographic sources, during the Twentieth Dynasty, the influence exerted on Egyptian defensive equipment by Western Asiatic helmets and body armour gradually weakened in favour of new types of coverings. The reign of Ramses III saw the widespread use of hooded helmets with two tassels at the top among Egyptian infantrymen, charioteers and 'marines'. Unfortunately, we do not have sufficient evidence to establish the actual nature of these headdresses and the crafting techniques employed. Nonetheless, their widespread use, particularly among Egyptian ships' soldiers – who also wore corslets apparently crafted in the same way – leads us to believe that the systematic adoption of heavier defensive equipment was a response to the challenges brought by new enemies. At the same time, a different type of helmet – a tight-fitting skullcap adorned with two tassels at the top – appears in several iconographic sources relating to the chariot equipment.

It is necessary to note here the ambiguous relationship between the sovereign of the New Kingdom and body armour. Although the use of corslets by pharaohs is amply attested, they failed to gain a foothold in the context of royal ideology, where the chariot and *khopesh* sickle-shaped sword best embodied the pharaoh's military virtues.

Finally, a perhaps distant but intriguing prospect would be the extension of research to the Egyptian religious sphere to recognize possible correlations between protective clothing and some divine figures. Indeed, I do not think it is incorrect to assume that armour could also find its place among the innumerable attributes associated with the powers and functions of Egyptian deities. The same reasoning can be applied to the warlike Asian gods whose worship spread through the Egyptian Empire in the same time frame that this military technology was introduced into the Egyptian army.

Appendix I

Helmets as depicted in Egyptian Art during the 18th Dynasty		
Reign	*Foreign Helmets*	*Egyptian Helmets*
Thutmose II (c. 1482 – 1480 BCE)		
Thutmose III (c. 1479 – 1425 BCE)		
Amenhotep II (c. 1425 – 1400 BCE)		
Thutmose IV (c. 1400 – 1390 BCE)		
Tutankhamun (c. ? – 1324 BCE)		
Aÿ (c. 1323 – 1320 BCE)		
Horemheb (c. 1319 – 1292)		

Appendix II

	Helmets as depicted in Egyptian Art during the Ramesside Period	
Reign	Foreign Helmets	Egyptian Helmets
Seti I (ca. 1290 – 1279 BCE)		
Ramesses II (ca. 1279 – 1213 BCE)		
Merenptah (ca. 1212 – 1202 BCE)		
Ramesses III (ca. 1187 – 1157 BCE)		

Appendix III

	Corslets as depicted in Egyptian Art during the New Kingdom	
Reign	*Foreign corslets*	*Egyptian corslets*
Ahmose I (ca. 1539 – 1515 BCE)		
Amenhotep II (ca. 1425 – 1400 BCE)		
Thutmose IV (ca. 1400 – 1390 BCE)		
Ramesses III (ca. 1187 – 1157 BCE)		

Notes

1 Introduction

1 J. Keegan, *A History of Warfare*, New York (1993), 27.
2 D. Dawson, *The First Armies*, London (2001), 15.
3 It is noteworthy that the Pharaonic armament reproduced on plate 88 of *Description de l'Égypte* probably provided the model for the weapons depicted in the allegorical scene *L'Étude et le Gènie dévoilent l'antique Égypte à la Grèce* painted in 1827 by François-Edouard Picot on the ceiling of the fourth room of the Musée Charles X (now Room XXX of the Louvre's Department of Egyptian antiquities).
4 C. L. Woolley, *The Royal Cemetery: A Report on the Predynastic and Sargonid Graves Excavated between 1926 and 1931* (2 vols), Ur Excavations, Vol. 2, Oxford (1934), 63–4, pl. 218 *a-b-c*; Y. Yadin, *The Art of Warfare in Biblical Lands in the Light of Archaeological Discovery*, New York (1963), 49; J. Aruz and R. Wallenfels, eds, *Art of the First Cities: The Third Millenium BC from the Mediterranean to the Indus*, New York (2003), 103; J.-L. Montero Fenollós, 'El armamento defensivo del soldado de Súmer y Mari', *AuOr* 21 (2003), 216, figure 4.3; W. J. Hamblin, *Warfare in the Ancient Near East to 1600 BC: Holy Warriors at the Dawn of History*, London (2006), 48–9; G. Gernez, *Les armes du Proche-Orient des origines à 2000 av. J.-C.*, Arles (2017), 96–7; C. Trimm, *Fighting for the King and the Gods: A Survey of Warfare in the Ancient Near East*, Atlanta (2017), 545.
5 T. Molleson and D. Hodgson, 'The Human Remains from Woolley's Excavations at Ur', *Iraq* 65 (2003), 107–11, fig. 13–14.
6 Following the discovery in 1903, Gaston Cros suggested a reconstruction drawing of the helmet (A. M. G. Cros, 'Mission Française de Chaldée: Campagne de 1903 Compte Rendu Sommaire des Fouilles', *RAAO* 6.1 (1904), 13, 16) which soon turned out to be incorrect. Only two years later, Cros proposed a new reconstruction (A. M. C. Cros, 'Note Rectificative: Sur le Casque Chaldéen de Tello: Lettre de M. le Commandant Gaston Cros', *RAAO* 6.3 (1906), 88–9), which is still reproduced in current academic publications. See A. Parrot, *Tello: Vingt campagnes de fouilles (1877–1933)*, Paris (1948), 106, fig. 26d and Montero Fenollós, *AuOr* 21 (2003), 215, figure 4.3.
7 C. L. Woolley, *The Royal Cemetery*, 60–2, pl. 90–3; Y. Yadin, *Art of Warfare*, 132–3; J. Aruz and R. Wallenfels, eds, *Art of the First Cities*, 97–100.

8 F. De Backer, *ResAnt* 8 (2011), 75.
9 Y. Yadin, *Art of Warfare*, 49; F. De Backer, *ResAnt* 8 (2011), 75, fig. 30; F. De Backer, 'Siege-Shield and Scale Armour Reciprocal Predominance and Common Evolution', *Historiae* 8 (2015), 2–3.
10 D. Howard, *Bronze Age Military Equipment*, Barnsley (2011), 72. Nevertheless, it should be said that the theory about cloaks made from animal fur had already been proposed by A. Hnila Gilibert in her article 'Warfare techniques in Early Dynastic Mesopotamia', *Proceeding of the International Symposium, Arms and Armour through the Ages, Modra-Harmónia, 19th–22nd November 2005* (*Anodos, Studies of the Ancient World* 4–5 / 2004–6), Trnva (2006), 98.
11 For a detailed overview of the discovery of the stele, its restoration and the studies linked to it, see L. Heuzey and F. Thureau-Dangin, *Restitution matérielle de la Stèle des Vautours*, Paris (1909); A. Parrot, *Tello*, 95–101, fig. 23; Y. Yadin, *Art of Warfare*, 134–5; M.-T. Barrelet, 'Peut-On Remettre en Question la "Restitution Matériel de la Stèle des Vautours"?', *JNES* 29.4 (1970), 233–58; I. J. Winter, 'After the Battle Is Over: The "Stele of the Vultures" and the Beginning of Historical Narrative in the Art of the Ancient Near East', *Symposium Papers IV: Pictorial Narrative in Antiquity and the Middle Ages* (Studies in the History of Art, Vol. 16), Washington (1985), 11–35; B. Alster, 'Images and Text on the "Stele of the Vultures"', *AOF* 50 (2003/4), 1–10; L. Romano, 'La Stele degli Avvoltoi: Una Rilettura Critica', *VicOr* XIII (2007), 3–23; I. Schrakamp, *Krieger und Waffen im frühen Mesopotamien*, Marburg (2010), 1–9.
12 Concerning the mass military formation adopted by the soldiers of Lagash, D. Nadali suggests that it could have been composed of two types of troops: soldiers who hold their long spears with both hands and shield-bearers in the front rank (D. Nadali, 'How Many Soldiers on the "Stele of the Vultures"? A Hypothetical Reconstruction', *Iraq* 76 (2014), 141–8).
13 See Aruz and R. Wallenfels, eds, *Art of the First Cities*, 90–2; D. Nadali, 'Monuments of War, War of Monuments: Some Considerations on Commemorating War in the Third Millennium BC', *Orientalia* 76, fasc. 4 (2007), 341–3, fig. 3.
14 A. Parrot, 'Les fouilles de Mari: Première campagne (hiver 1933–1934)', *Syria* 16, fasc. 2 (1935), 132–7, pl. XXVIII; *idem*, *Mission archéologique de Mari*. Vol. I: *le temple d'Ishtar*, Paris (1956), 136–55, pl. LVI–LVII; Y. Yadin, *Art of Warfare*, 137–9; R. Dolce, *Gli intarsi mesopotamici dell'epoca protodinastica*, Rome (1978), 134–6, pl. XXXVII; J.-L. Montero Fenollós, *AuOr* 21, 219–20, fig. 3: 2–3; D. Nadali, *Orientalia* 76, fasc. 4, 343–5, fig. 5; P. Butterlin, 'Mari et l'histoire militaire mésopotamienne: du temps long au temps politico-militaire', in M. d'Andrea *et al.*, eds, *Pearls of the Past, Studies on Near Eastern Art and*

Archaeology in Honour of Frances Pinnock (*marru* 8), Münster (2019), 14–15, fig. 11.

15 R. Dolce, *Gli intarsi*, 149, 152, pl. XXXVI; J. Aruz and R. Wallenfels, eds, *Art of the First Cities*, 157–8; J.-C. Margueron, *Mari: Métropole de l'Euphrate, au IIIe millénaire av. J.-C.*, Paris (2004), 197–207, 289–91; J.-L. Montero Fenollós, *AuOr* 21, 219, fig. 3: 4; D. Nadali, *Orientalia* 76, fasc. 4, 346, fig. 5; P. Butterlin, 'Mari et l'histoire militaire mésopotamienne: du temps long au temps politico-militaire', 14, fig. 12.

16 Aruz and R. Wallenfels, eds, *Art of the First Cities*, 175–7; Nadali, *Orientalia* 76, fasc. 4, 348–51; G. Minunno, 'Pratiche di mutilazione dei nemici caduti nel Vicino Oriente Antico', *Mesopotamia* XLIII (2008), 10–11, fig. 1; P. Matthiae, 'The Victory Panel of Early Syrian Ebla: Finding, Structure, Dating', *Studia Eblaitica* 3 (2017), 33–83.

17 W. J. Hamblin, *Warfare*, 247; P. Butterlin, 'Mari et l'histoire militaire mésopotamienne: du temps long au temps politico-militaire', 14.

18 A. Parrot, 'Les fouilles de Mari: Dix-neuvième campagne (printemps 1971)', *Syria* 48, fasc. 3–4 (1971), 269, pl. XIV, 4; J. Aruz and R. Wallenfels, eds, *Art of the First Cities*, 158–9, n. 99; J.-L. Montero Fenollós, *AuOr* 21, 218–19, fig. 3: 1; P. Butterlin, 'Mari et l'histoire militaire mésopotamienne: du temps long au temps politico-militaire', 9, fig. 7.

19 Deir Ez-Zor has been the scene of heavy fighting during the Syrian civil war. The threat posed by Daesh in the area forced the Syrian army to move the collection to a safer place in 2015. At present we are unable to determine the actual position of the archaeological artefacts.

20 For further details about the shield and the bow depicted on the stone, see Y. Yadin, 'The Earliest Representation of a Siege Scene and a "Scythian Bow" from Mari', *IEJ* 22 (1972), 89–94; W. J. Hamblin, *Warfare*, 89–90, 218, fig. 5c; F. De Backer, 'Les archers de siège néo-assyriens' in G. Wilhelm, ed., *Organization, Representation, and Symbols of Power in the Ancient Near East: Proceedings of the 54th Rencontre Assyriologique Internationale at Würzburg, 20–25 July 2008*, Winona Lake, IN (2012), 433, 439–40, fig. 16.

21 The Akkadian term *tutittu* generally designates a sort of breastplate which can be considered an early form of *kardyophylax*, a chest armour, primarily protecting the area around the heart, commonly used in Anatolia and Western Asia during the first millennium BCE. See T. Kendall, 'gurpisu ša awēli: The Helmets of the Warriors of Nuzi' in M. A. Morrison, D. I. Owen, eds, *Studies on the Civilization and Culture of Nuzi and the Hurrians*, Vol. 1: *in Honor of Ernst R. Lacherman*, Winona Lake, IN (1981), 202–3, n. 9; F. De Backer, *Historiae* 8, 3. In this respect, it is worth recalling that *kardyophylakes*/cuirass discs were widespread among Italic,

Celtic and Iberian peoples between the eighth and the second century BCE. The bibliography concerning the *kardyophylax* is too broad to be manageable. We nevertheless suggest the following further readings that we see as a starting point for deepening this topic: R. Papi, 'Produzione metallurgica e mobilità nel mondo italico' in L. Del Tutto Palma, ed., *La Tavola di Agnone nel contesto italico: Atti del convegno di studio. Agnone 13–15 aprile 1994*, Florence (1996), 89–128; G. Tomedi, *Italische Panzerplatten und Panzerscheiben*, Stuttgart (2000); A. Cherici, 'Sulle rive del Mediterraneo centro-occidentale: aspetti della circolazione di armi, mercenari e culture' in G. M. Della Fina, ed., *Etruschi Greci Fenici e Cartaginesi: Atti del XIV Convegno Internazionale di Studi sulla Storia e l'Archeologia dell'Etruria (2006)* (Annali della Fondazione per il Museo Claudio Faina, Vol. XIV), Rome (2007), 235–46; G. Colonna, 'Dischi-corazza e dischi

di ornamento femminile: due distinte classi di bronzi centro-italici', *ArchClass* 58 (2007), 3–30; F. De Backer, 'Cardiophylax en Urartu: un modèle celtibère' in P. Bieliński *et al.*, eds, *Proceedings of the 8th International Congress on the Archaeology of the Ancient Near East: 30 April–4 May 2012, University of Warsaw.* Vol. I, Wiesbaden (2014), 569–84; R. Papi, 'Guerrieri di pietra e dischi di bronzo', *Picus* 41 (2021), 9–84.

22 For an overview of the Akkadian army, see P. Abrahami, 'L'armée d'Akkad' in P. Abrahami and L. Battini, eds, *Les armées du Proche-Orient ancien: IIIe–Ier mill. av. J. C. Actes du Colloque International Organisé à Lyon les 1er et 2 décembre 2006, Maison de l'Orient et de la Méditerranée*, Oxford (2008), 1–20.

23 Two large fragments are on display at the Louvre Museum (AO 2678, AO 2679); the third fragment belongs to the Yale Babylonian Collection (Inv. N. 2409). See A. Parrot, *Tello*, 133–4, pl. Xb; P. Amiet, *L'art de l'Agadé au musée du Louvre*, Paris (1976), 26–7; B. R. Foster, 'The Sargonic Victory Stele from Telloh', *Iraq* 47 (1985), 15–30; L. Nigro, 'La stele di Rimush da Tello e l'indicazione del rango dei vinti nel rilievo reale accadico', *Scienze dell'Antichità* 11 (2001–3), 71–6, fig. 1–3; J. Aruz and R. Wallenfels, eds, *Art of the First Cities*, 201–2; W. J. Hamblin, *Warfare*, 86; A. Thomas and T. Potts, eds, *Mesopotamia: Civilisation Begins*, Los Angeles (2020), 179.

24 The Victory stele of Naram-Sin was brought from Sippar (modern Tell Abu Habbah, Iraq) to Susa as spoil of war by Shutruk-Nakhunte in the twelfth century BCE. The Elamite king added an inscription to the stele celebrating the defeat of Sippar. The stele can be seen at Louvre Museum (Sb 4). See M.-T. Barrelet, 'Notes sur quelques sculptures mésopotamiennes de l'époque d'Akkad', *Syria* 36, fasc. 1–2 (1959), 31–3, fig. 5; P. Amiet, *L'art de l'Agadé*, 29–32; J. Aruz and R. Wallenfels, eds, *Art of the First Cities*, 195–7, fig. 59; W. J. Hamblin, *Warfare*, 86, fig. 3; A. Thomas and T. Potts, eds, *Mesopotamia*, 67–8, fig. 66.

25 F. Basmachi, 'An Akkadian Stele', *Sumer* X (1954), 116-19; J. F. X. McKeon, 'An Akkadian Victory Stele', *BMFA* 68.354 (1970), 226-43; J. Aruz and R. Wallenfels, eds, *Art of the First Cities*, 204; W. J. Hamblin, *Warfare*, 87; G. Gernez, *Les armes*, 121.

26 Stylistic similarities and the use of the same type of stone have led some scholars to hypothesize that the fragmentary stele from Nasiriyah (Museum of Fine Arts of Boston MFA 66.893; Iraq Museum of Baghdad (IM 55639, IM 59205) and the two fragments from Susa (Louvre Museum Sb 6641, Sb 6641 bis) have been made in the same workshop. It is reasonable to think that the fragments excavated at Susa were part of a stele looted from Mesopotamia by the Elamite king Shutruk-Nakhunte (P. Amiet, *L'art de l'Agadé*, 243; J. F. X. McKeon, *BMFA* 68.354, 237-8, fig. 18).

27 The well-known models of soldiers from the tomb of Mesehti are currently kept in the Cairo Museum (CGC 257, 258). See L. Borchardt, *Statuen und Statuetten von Königen und Privatleuten im Museum von Kairo, Nr. 1-1294. Teil I., Text und Tafeln zu Nr. 1-380*, Berlin (1911), 164-5, pl. 55-6; W. Wolf, *Die Bewaffnung des altägyptischen Heer*, Leipzig (1926), 103, pl. 20; Y. Yadin, *Art of Warfare*, 163.

28 D. Howard, *Bronze Age Military Equipment*, 71.

29 See H. O. Willems, 'The Nomarchs of the Hare Nome and Early Middle Kingdom History', *JEOL* 28 (1983-4), 83-4; P. Tallet, *Sésostris III et la fin de la XIIe dynastie*, Paris (2015), 87.

30 F. Monnier, 'La scène de traction du colosse de Djéhoutyhotep: Description, traduction et reconstitution', *JAEA* 4 (2020), 55-72 represents the most complete and updated study concerning this well-known scene.

31 See P. Newberry, *El Bersheh (Band I): The Tomb of Tehuti-Hetep*, London (1985), 32-9, pl. XXIV, XXIX.

32 Ibid, 17, pl. XIII.

33 B. McDermott, *Warfare in Ancient Egypt*, Gloucestershire (2004), 52, fig. 34.

34 See above n. 20. The bronze ellipsoidal *kardiophylax* from Carpena, near Forlì, Emilia-Romagna, Italy, dating back to the seventh century BCE is an excellent example of this type of breastplate. See G. Bermond Montanari, 'La tomba di Carpena (Forlì)' in G. Bermond Montanari *et al.*, eds, *Quando Forlì non c'era: Catalogo della mostra*, Forlì (1997), 273-7; L. Malnati, 'Armi e organizzazione militare in Etruria padana' in G. M. Della Fina, ed., *La colonizzazione etrusca in Italia: Atti del XV Convegno Internazionale di Studi sulla Storia e l'Archeologia dell'Etruria (2007)*, (Annali della Fondazione per il Museo Claudio Faina, Vol. XV), Rome (2008), 151, n. 19, pl. 6, 1.

35 Although Sydney Aufrère claims that the human remains have been found in tombs MMA 101, 506 and 507 at Deir el-Bahari (S. Aufrère, 'Les vétérans de

Montouhotep Nebhépetrê. Une garnison funéraire à Dei el-Bahari?' *Egypte* 19 (2000), 10), they were discovered only in tomb MMA 507. This incongruity was first noticed by Carola Vogel (C. Vogel, 'Fallen Heroes? – Winlock's 'Slain Soldiers' Reconsidered', *JEA* 89 (2003), 241).

36 See H. E. Winlock, 'The Egyptian Expedition 1925–1927: The Museum's Excavations at Thebes', *BMMA* 23 (1928), 12; idem, *Excavations at Deir el Baḥri 1911–1913*, New York (1942), 123; idem, *The Slain Soldiers of Neb-ḥepet-Rē' Mentu-ḥotpe*, New York (1945), 1–2; S. Aufrère, *Egypte* 19, 10; C. Vogel, *JEA* 89, 240. The abandoned mortuary temple of Hatshepsut (*c*. 1478–1458 BCE) at Deir el-Bahari was the place where Apa Abraham, the fourteenth bishop of the diocese of Hermontis, decided to found a monastery dedicated to St Phoibammon in the sixth century CE. Today only a few traces of the monastery remains on the upper terrace of Hatshepsut's temple (W. Godlewski, 'Remarques sur la creation du monastère de St. Phoibammon à Deir el-Bahari', *AfrBull* 31 (1982), 107–14; idem, *Deir El-Bahari V: Le monastère de Phoibammon*, Warsaw (1986), 60–78; F. Krueger, 'Revisiting the First Monastery of Apa Phoibammon: A Prosopography and Relative Chronology of Its Connections to the Monastery of Apa Ezekiel within the Monastic Network of Hermonthis during the Sixth Century', *APF* 66.1 (2020), 151–2).

37 H. E. Winlock, *BMMA* 23, 12, fig. 13; idem, *Excavations*, 123, pl. 21; idem, 'The Eleventh Egyptian Dynasty', *JNES* 2 (1943), 275–6; idem, *Slain Soldiers*, 25–31, pl. XVI–XX.

38 H. E. Winlock, *Excavations*, 127, pl. 20; idem, *Slain Soldiers*, 10, pl. IV–V; S. Aufrère, *Egypte* 10, 10, fig. 3–4; R. B. Partridge, *Fighting Pharaohs. Weapons and Warfare in Ancient Egypt*, Manchester (2002), 120–2; C. Vogel, *JEA* 89, 240.

39 H. E. Winlock, *BMMA* 23, 13–15, fig. 16–19; idem, *Excavations*, 124–7, pl. 19; idem, *Slain Soldiers*, 7–24; S. Ikram and A. Dodson, *The Mummy in Ancient Egypt: Equipping the Dead for Eternity*, London (1998), 116; S. Aufrère, *Egypte* 10, 10–11; G. M. Sanchez, 'A Neurosurgeon's View of the Battle of Reliefs of King Sety I: Aspect of Neurological Importance', *JARCE* 37 (2000), 162, 164; C. Vogel, *JEA* 89, 239; H. Köpp-Junk, 'Quellen zum Krieg im alten Ägypten' in H. Meller and M. Schefzik, eds, *Krieg – eine archäologische Spurensuche*, Halle (2015), 231.

40 H. E. Winlock, *Slain Soldiers*, 9; J. Fletcher, 'Ancient Egyptian Hair and Wigs', *The Ostracon: The Journal of the Egyptian Study Society* 13.2 (2002), 3; B. McDermott, *Warfare*, 50.

41 H. E. Winlock, *BMMA* 23, 16; idem, *Excavations*, 125–6; idem, *Slain Soldiers*, 8, 24; S. Aufrère, *Egypte* 10, 14.

42 D. Arnold, *Der Temple des König Mentuhotep in Deir el-Bahari, Band* III: *Die königlichen Beigaben*, Mainz (1981), 47–8.

43 The relief fragments have been published for the first time in E. Neville, *The XIth Dynasty Temple at Deir el-Bahari*, Vol. I, London (1907), 68–9, pl. XIV–XV.
44 This reconstruction is supported by many studies: S. W. Smith, *Interconnections in the Ancient Near East: A Study of the Relationships between the Arts of Egypt, the Aegean, and Western Asia*, New Haven, CN, London (1965), 148–9, fig. 185, pl. XIV; G. A. Gaballa, *Narrative in Egyptian Art*, Mainz (1976), 36–7, fig. 2b; C. Vogel, *Ägyptische Festungen und Garnisonen bis zum Ende des Mittleren Reiches*, Hildesheim (2004), 54, fig. 9; A. M. Pollastrini, 'La poliorcetica in Egitto dall'Antico Regno alla XXV dinastia' in P. Gallo, ed., *Egittologia a Palazzo Nuovo: Studi e ricerche dell'Università di Torino*, Novi Ligure (2013), 240, pl. IV, fig. 9; F. Monnier, 'Une iconographie égyptienne de l'architecture défensive', *ENiM* 7 (2014), 178–80, fig. 5; L. Bestock, *Violence and Power in Ancient Egypt: Image and Ideology before the New Kingdom*, London, New York (2018), 127–39, fig. 4.34–4.44 ; F. Monnier, 'Les techniques de siège décrites dans la documentation pharaonique', *ENiM* 15 (2022), 55, n. 18.
45 S. Aufrère, *Egypte* 10, 16.
46 See B. Jaroš-Deckert, *Grabung im Asasif 1963–1970. Band V. 'Das Grab des 'Ini-iti.f'. Die Wandmalereien der 11. Dynastie*, Mainz (1984), pl. 1; C. Vogel, *Ägyptische Festungen*, 50–4, fig. 8; A. M. Pollastrini, 'La poliorcetica in Egitto dall'Antico Regno alla XXV dinastia', 239; F. Monnier, *ENiM* 7, 175–6, 196; L. Bestock, *Violence and Power*, 235–41, figure 8.7–8.8; F. Monnier, *ENiM* 15, 57 – fig. For a reconstruction of the wheeled siege tower shown in the tomb of Intef, see F. Monnier, 'Proposition de reconstitution d'une tour de siège de la XIe dynastie', *JSSEA* XXXIX (2012–13), 125–38.
47 C. Vogel, *JEA* 89, 244–5.

2 Reception and Diffusion of Personal Protections during the Eighteenth Dynasty

1 M. A. Littauer and J. H. Crouwel, *Wheeled Vehicles and Ridden Animals in the Ancient Near East*, Leiden (1979) is always a reference text. To this must be added A. J. Veldmeijer and S. Ikram, eds, *Chasing Chariots: Proceedings of the First Chariot Conference* (Cairo 2012), Leiden (2013) and A. J. Veldmeijer and S. Ikram, eds, *Chariots in Ancient Egypt: The Tano Chariot, a Case of Study*, Leiden (2018).
2 In the text of the Second Kamose Stele, the word transliteration probably refers to a team of horses yoked to a chariot. See L. Habachi, *The Second Stele of Kamose and His Struggle against the Hyksos Ruler and His Capital*, Glückstadt (1972), 36 and

O. Goldwasser, 'What is a Horse? Lexical Acculturation and Classification in Egyptian, Sumerian, and Nahuatl' in T. Pommerening and W. Bisang, eds, *Classification from Antiquity to Modern Times: Sources, Methods, and Theories from an Interdisciplinary Perspective*, Berlin, Boston (2017), 49, 52.

3   As regards the rock-cut tomb of Ahmose, son of Abana at Elkab (PM V, 182) see W. V. Davies, 'The Tomb of Ahmose Son-of-Ibana at Elkab. Documenting the Family and Other Observations' in W. Claes, H. de Meulenaere and S. Hendrickx, eds, *Elkab and Beyond. Studies in Honour of Luc Limme*, Leuven (2009), 139–76.

4   The emergence of the chariot has made necessary a change in the method of how to arrange troops on the battlefield. See A. J. Spalinger, 'Le positionnement des troupes sur le champs de bataille du Nouvel Empire', *Egypte* 106 (2022), 3–16.

5   See C. Trimm, *Fighting*, 203–18.

6   The growing importance of horsemanship was probably the main reason for the premature decline of chariot tactics in the Ancient world. In this respect, see R. Archer, 'Chariotry to Cavalry: Developments in the Early First Millennium' in G. G. Fagan and M. Trundle, eds, *New Perspectives on Ancient Warfare*, Leiden, Boston (2010), 66–73.

7   The Akkadian origin of the composite bow during the third millennium BCE is still debated. See G. Gernez, *Les armes*, 122–4.

8   For further information regarding the origin and the development of the sickle-shaped sword, see H. W. Müller, *Der Waffenfund von Balâṭa-Sichem und Die Sichelschwerter*, Munich (1987), 108–76; C. Vogel, 'Hieb- und stichfest? Überlegungen zur Typologie des Sichelschwertes im Neuen Reich' in D. Bröckelmann and A. Klug, eds, *In Pharaos Staat: Festschrift für Rolf Gundlach zum 75. Geburtstag*, Wiesbaden (2006), 271–86; G. Gernez, 'Des armes et des hommes: La question des modèles de diffusion des armes au Proche-Orient à l'Âge du Bronze' in P. Rouillard, ed., *Mobilités Immobilismes: L'emprunte et son refus*, Paris (2007), 128–30; A. Massafra, *Le harpai nel Vicino Oriente antico: cronologia e distribuzione*, Rome (2012), 11–45; N. Couton-Perche, *Les armes de l'Égypte ancienne: La collection du musée du Louvre*, Paris (2021), 236–43.

9   For an overview of the characteristics of this type of dagger, see H. W. Müller, *Der Waffenfund*, 61–7; A.-L. Mourad, 'Transforming Egypt into the New Kingdom: The Movement of Ideas and Technology across Geopolitical, Cultural and Social Borders' in M. Bietak and S. Prell, eds, *The Enigma of the Hyksos*, Vol. IV: *Changing Clusters and Migrations in the Near Eastern Bronze Age*, Wiesbaden (2021), 465–7, fig. 8. Unlike the sickle-shaped sword, the dagger with the blade and handle cast in a single piece crossed the boundaries of the Near East. In this connection, see T. J. Papadopoulos, *The Late Bronze Age Daggers of the Aegean*, Vol. 1: *The Greek Mainland*, Stuttgart (1998), 16–17, pl. 10–11.

10 See Y. Yadin, *Art of Warfare*, 84; R. Drews, *The End of the Bronze Age. Changes in Warfare and the Catastrophe ca. 1200 BC*, Princeton, NJ (1993), 319; P. R. S. Moorey, 'The Mobility of Artisans and Opportunities for Technology Transfer between Western Asia and Egypt in the Late Bronze Age' in A. J. Shortland, ed., *The Social Context of Technological Change: Egypt and the Near East, 1650-1150 BC. Proceedings of a Conference Held at St Edmund Hall, Oxford 12-14 September 2000*, Oxford (2001), 8; T. Dezsö, 'Scale Armour of the 2nd Millennium BC' in T. A. Bács, ed., *A Tribute to Excellence: Studies Offered in Honor of Ernö Gaal, Ulrich Luft, László Török*, Budapest (2002), 195; idem, 'Panzer' in E. Ebeling, E. Weidner, eds, *RLA*, Band 10 (2003-5), 319.

11 G. Wilhelm, *The Hurrians*, Warminster (1989), 17-19; S. Turner, *The Horse in The New Kingdom Egypt: Its Introduction, Nature, Role and Impact*, Wallasey (2021).

12 P. Raulwing, 'The Kikkuli Text (CTH 284): Some Interdisciplinary Remarks on Hittite Training Texts for Chariot Horses in the Second Half of the 2nd Millennium BC' in A. Gardeisen, ed., *Les Équidés dans le monde méditerranéen antique: Actes du colloque organise par l'École française d'Athènes, le Centre Camille Jullian et l'UMR 5140 du CNRS, Athèns, 26-28 novembre 2003*, Lattes (2005), 61-75.

13 *Maryannu* is a word of Indo-Iranian origin and is probably connected with the Sanskrit term *marya*, 'young warrior'. See G. Wilhelm, *The Hurrians*, 19; R. Drews, *The Coming of the Greeks: Indo-European Conquests in the Aegean and the Near East*, Princeton, NJ (1988), 59-60; J. E. Hoch, *Semitic Words in Egyptian Texts f the New Kingdom and Third Intermediate Period*, Princeton, NJ (2004), 135-7 [175].

14 For an overview of the characteristics of the *maryannu* chariot warriors, see W. F. Albright, 'Mitannian maryannu "chariot-warrior", and the Canaanite and Egyptian Equivalent', *AOF* 6 (1930-1), 217-21; R. T. O'Callaghan, 'New Light on the Maryannu as "Chariot Warrior"', *JKAF* 1 (1951), 309-24; H. Reviv, 'Some comments on the Maryannu', *IEJ* 22 (1972), 218-28; M. R. Abbas, 'The Maryannu in the Western Desert during the Ramesside Period', *Abgadiyat* 8 (2013), 127-32.

15 See J. C. Darnell and C. Manassa, *Tutankhamun's Armies: Battle and Conquest during Ancient Egypt's Late 18th Dynasty*, Hoboken, NJ (2007), 64-5. Moreover, the word *maryannu* became a title within the Egyptian army. See, for example, A. F. Rainey, 'The Soldier-Scribe in Papyrus Anastasi I', *JNES* 26 (1967), 58-60.

16 R. de Vaux, 'Les Ḫurrites de l'Histoire et les Horites de la Bible', *CRAI(BL)* 111e année/3, 428; P. R. S. Moorey, 'The Mobility of Artisans and Opportunities for Technology Transfer between Western Asia and Egypt in the Late Bronze Age', in A. J. Shortland, ed., *The Social Context of Technological Change: Egypt and the Near

East, 1650-1150 BC. Proceedings of a Conference Held at St Edmund Hall, Oxford 12-14 September 2000, Oxford (2001), 6-9.
17 See below section 5.3 𓃀𓏏𓇋𓈖𓏥 *tryn*.
18 V. D. Hanson, *The Western Way of War: Infantry Battle in Classical Greece*, New York (1989), 71-83.
19 See S. Harvey, 'Monuments of Ahmose at Abydos', *EgArch* 4 (1994), 3-5; idem, 'Abydos' in G. J. Stein, ed., *The Oriental Institute 2002-2003 Annual Report*, Chicago (2003), fig. 8; idem, 'New Evidence at Abydos for Ahmose's funerary cult', *EgArch* 24 (2004), 3-6. See also discussion in: J. Bourriau, 'The Second Intermediate Period (*c*. 1650-1550 BC)' in I. Shaw, ed., *The Oxford History of Ancient Egypt*, Oxford (2000), 213, figure on p. 213; A. J. Spalinger, *War in Ancient Egypt. The New Kingdom*, Oxford (2005), 19-22, figure 1.4-1.7; C. Barbotin, *Âhmosis et le début de la XVIIIe dynastie*, Paris (2008), 121-2; fig. 32; R. Pietri, 'Le roi en char au Nouvel Empire', *Egypte* 74 (2014), 13-14, fig. 1; J. Iwaszczuk, 'Battle Scenes from the Temple of Thutmose I in Qurna', *ÄgLev* 31 (2021), 157.
20 For more information on memorial monuments built under the reigns of Hatshepsut and Thutmose III on the West Bank, see J. Iwaszczuk, *Sacred Landscape of Thebes during the Reign of Hatshepsut. Royal Construction Project.* Vol. I: *Topography of the West Bank*, Warsaw (2017); C. Karlshausen and T. De Putter, 'From Limestone to Sandstone – Building Stone of Theban Architecture During the Reigns of Hatshepsut and Thutmosis III', *JEA* 106 (1-2) (2020), 221-4.
21 B. Bruyère, *Deir el Médineh. Année 1926. Sondage au temple funéraire de Thotmès II (Hat Ankh Shesept)*, Cairo (1952), 40-2, pl. III-IV. See also discussion in: A. R. Schulman, 'Hittites, Helmets and Amarna: Akhenaten's First Hittite War', in D. Redford, ed., *The Akhenaten Temple Project*, Vol. 2: *Rwd - mnw, Foreigners and Inscriptions*, Toronto (1988), 71, n. 27; A. J. Spalinger, *War in Ancient Egypt*, 60, fig. 31a and b; J. Iwaszczuk, *ÄgLev* 31, 157.
22 L. Gabolde and M. Gabolde, 'Les temples 'mémoriaux' de Thoutmosis II et Toutânkhamon (un rituel destine à des statues sur barques)', *BIFAO* 89 (1990), 128-39; L. Gabolde, *Monuments décorés en bas relief aux noms de Thoutmosis II et Hatchepsout à Karnak*, Cairo (2005), 175-6; R. Pietri, *Egypte* 74, 14.
23 The military campaign undertaken by Thutmose II to punish the Shasu nomads is mentioned in the well-known Autobiography of Ahmose Pen-Nekhebet (*Urk* IV, 36, 12-14), a soldier who served Ahmose I, Amenhotep I, Thutmose I, Thutmose II, Hatshepsut and Thutmose III. See also discussion in: R. Giveon, *Les Bédouins Shosou des documents égyptiens*, Leiden (1971), 9-10; E. F. Morris, *The Architecture of Imperialism: Military Bases and the Evolution of Foreign Policy in Egypt's New Kingdom*, Leiden, Boston (2005), 33.

The text of the Autobiography (*Urk* IV, 32–39) decorates the entrance wall of Ahmose Pen-Nekhebet's rock tomb in El-Kab (PM V, 176–177) as well as the surface of two fragmentary statues (PM V, 177, 191), one now in the Louvre Museum (C 49) and the other in the National Museum of Scotland, Edinburgh (NMS 1848.486). See, W. V. Davies, 'A View from Elkab: The Tomb and Statues of Ahmose-Pennekhebet' in J. M. Galán *et al.*, eds, *Creativity and Innovation in the Reign of Hatshepsut*, Chicago (2014), 401–7.

24 In the middle colonnade of the mortuary temple of Hatshepsut, nine fragmentary columns of hieroglyphic text, associated with a relief representing tributes or war prizes, probably allude to a military campaign of Thutmose II in Retjenu. See, K. Sethe, *Untersuchungen zur Geschichte und Altertumskunde Aegyptens*, Leipzig (1896), 40; E. Naville, *The Temple of Deir el Bahari. Part III: End of Northern Half and Southern Half of the Middle Platform*, London (1898), 17, pl. LXXX; J. H. Breasted, *Historical Records of Egypt*. Vol. II: *The Eighteenth Dynasty*, Chicago (1906), 51.

More recently, however, the reign of Thutmose II has been identified as a period of relative peace without any relevant conflicts against the kingdom of Mitanni (D. B. Redford, *Egypt, Canaan and Israel in Ancient Times*, Princeton (1992), 154, n. 122; P. Grandet, *Les Pharaons du Nouvel Empire: une pensée stratégique (1550–1069 avant J.-C.)*, Monaco (2008), 76).

25 L. D. Morenz, 'Reconsidering Sheshonk's emblematic list and his war in Palestine' in P. Kousoulis and K. Magliveras, eds, *Moving across the Borders: Foreign Relations, Religion and Cultural Interactions in the Ancient Mediterranean*, Leiden (2007), 106–7; A. M. Pollastrini, 'Le armi dei popoli vinti nell'iconografia egiziana' in S. Graziani and G. Lacerenza, eds, *Egitto e Vicino Oriente antico tra passato e future. The Stream of Tradition: la genesi e il perpetuarsi delle tradizioni in Egitto e nel Vicino Oriente antico*, Naples (2022), 308–9.

26 PM I/2, 560.

27 H. Carter and P. E. Newberry, *The Tomb of Thoutmôsis IV*, Westminster (1904), 26–32. For a general description of the chariot body, see B. M. Bryan, *The Reign of Thutmose IV*, Baltimore and London (1991), 193–4; A. M. Calvert, 'Vehicle of the Sun: The Royal Chariot in the New Kingdom' in A. J. Veldmeijer and S. Ikram, eds, *Chariots in Ancient Egypt: The Tano Chariot, a Case of Study*, Leiden (2018), 46–56.

28 W. Wreszinski, *Atlas zur altägyptischen Kulturgeschichte*. II. Teil, Leipzig (1935), pl. 1.

29 See W. Helck, *Die Beziehungen Ägyptens zu Vorderasien im 3. und 2. Jahrtausend v. Ch*, 2nd ed., Wiesbaden (1971), 330–2; A. R. Schulman, 'Hittites, Helmets and Amarna', 54, n. 17; J. C. Darnell, 'Supposed Depictions of Hittites in the Amarna

Period', *SAK* 18 (1991), 128; J. C. Darnell and C. Manassa, *Tutankhamun's Armies*, 179–80.

30 W. Wreszinski, *Atlas* II, pl. 2.
31 Y. Yadin, *Art of Warfare*, 12; W. V. Davies, *Catalogue of Egyptian Antiquities in the British Museum* VII. *Tools and Weapons* I. *Axes*, London (1987), 43–6, pl. 18–23; N. Couton-Perche, *Les armes*, 80–2.
32 T. Kendall, 'gurpisu ša awēli: The Helmets of the Warriors of Nuzi' in M. A. Morrison, D. I. Owen, eds, *Studies on the Civilization and Culture of Nuzi and the Hurrians*, Vol. 1: *in Honor of Ernst R. Lacherman*, Winona Lake, IN (1981), 221–4.
33 See, eg, J. R. Zorn, 'Reconsidering Goliath: An Iron Age I Philistine Chariot Warrior', *BASOR* 360 (2010), 3, fig. 2.
34 According to the Bible, king Ahab of Israel (*c*. 871–852 BCE) died in the same way: 'Now a certain man drew a bow at random, and struck the king of Israel between the joints of his armor. So he said to the driver of his chariot, "Turn around and take me out of the battle, for I am wounded." The battle raged throughout that day, and the king was propped up in his chariot facing the Arameans. And the blood from his wound ran out onto the floor of the chariot, and that evening he died' (1 Kings 22.34–35). See also Y. Yadin, *Art of Warfare*, 196.
35 The 'Mansion of NebkheperuRe at Thebes' should not be confused with another building bearing the name of King Tutankhamun, the 𓉗𓉐 *Ḥwt Nb-Khpr.w-R^c mry 'Imn grg(w) W^{c}st* 'Mansion of Nebkheperure, Beloved of Amon, who puts Thebes in Order'. In this regard, see: R. Sa'ad, 'Fragments d'un monument de Toutânkhamon retrouvés dans le IXe pylône de Karnak', *CahKarn* 5 (1975), 93–109, pl. XXXIV–XXXVI; M. Eaton-Krauss, 'Tutankhamun in Karnak', *MDAIK* 44 (1988), 1–3.
36 According to Otto Schaden, the construction of the building started during the reign of Tutankhamun and continued under the reign of Ay. See O. J. Shaden, 'Tutankhamun and Ay blocks from Karnak', *NARCE* 80 (1972), 39–40; *idem*, 'Report on the 1978 Season at Karnak', *NARCE* 127 (1984), 44–64; *idem*, 'A Tutankhamun Stela at Karnak', *CahKarn* 8 (1978), 279–84; *idem*, 'Tutankhamon-Ay Shrine at Karnak and Western Valley of the Kings Project: Report on the 1985–1986 Season', *NARCE* 138 (1978), 10–15. Luc and Marc Gabolde suggest instead that Ay built the monument to honour the memory of his predecessor. See L. Gabolde and M. Gabolde, *BIFAO* 89, 139–41; M. Gabolde, *Toutankhamon*, 409–13. In this regard, see also M. Eaton-Krauss, *MDAIK* 44, 3–11; W. J. Murnane, *The Road to Kadesh: A Historical Interpretation of the Battle Reliefs of King Sety I at Karnak*, Chicago (1990), 18–19, n. 98.

37 According to W. Raymond Johnson, the *Ḥwt Nb-Khpr.w-Rꜥ m Wꜥst* probably stood on the West Bank of the Nile, opposite Thebes (personal communication, 3 February 2018). Marc Gabolde proposed instead that the memorial temple was built inside the Precinct of Amun-Re at Karnak. See M. Gabolde, *Toutankhamon*, Paris (2015), 409.
38 M. Gabolde, 'Horemheb et les campagnes égyptiennes en Asie sous Toutânkhamon et Aÿ', *Egypte* 76 (2014/15), 28.
39 A. Dodson, *Amarna Sunset: Nefertiti, Tutankhamun, Ay, Horemheb and the Egyptian Counter-Reformation*, Cairo, New York (2009), 66–8.
40 Latest research findings about the 'Mansion of Nebkheperure at Thebes' have been excellently summarized in M. Gabolde, *Toutankhamon*, 409–25, fig. 178–90. See also L. Bell, 'The Epigraphic Survey', *OIR 1986–87* (1987), 4–6; *idem*, 'The Epigraphic Survey', *OIR 1988–89* (1990), 4; W. R. Johnson, *An Asiatic Battle Scene of Tutankhamun from Thebes: A Late Amarna Antecedent of the Ramesside Battle-Narrative Tradition*, PhD diss., The University of Chicago, Chicago (1992); *idem*, 'Tutankhamen-period Battle Narratives at Luxor', *KMT* 20.4 (2009–10), 20–33.
41 A. J. Spalinger, *Icon of Power: A Strategy of Reinterpretation*, Prague (2011), 24.
42 See W. J. Murnane, *The Road to Kadesh*, 22; W. R. Johnson, *An Asiatic Battle Scene of Tutankhamun from Thebes*, 48–82; J. Martínez Babón, *Historia Militar de Egipto durante la Dinastía XVIII*, Barcelona (2003), 96–7; J. C. Darnell and C. Manassa, *Tutankhamun's Armies*, 178–82; P. Grandet, *Les Pharaons*, 164–5; A. Dodson, *Amarna Sunset*, 57.
43 V. Cordani, 'Suppiluliuma in Syria after the First Syrian War: The (Non-)Evidence of the Amarna Letters' in S. de Martino and J. L. Miller, eds, *New Results and New Questions on the Reign Suppiluliuma I*, Florence (2013), 43–64.
44 W. R. Johnson, *An Asiatic Battle Scene of Tutankhamun from Thebes*, 13, 165–8 [44–8]; J. Martínez Babón, *Historia Militar*, 96; J. C. Darnell and C. Manassa, *Tutankhamun's Armies*, 119–21; A. Dodson, *Amarna Sunset*, 68.
45 W. R. Johnson, *An Asiatic Battle Scene of Tutankhamun from Thebes*, 60–1.
46 A. R. Schulman, 'Hittites, Helmets and Amarna', 63, fig. 17; W. R. Johnson, *An Asiatic Battle Scene of Tutankhamun from Thebes*, 157–8, [17].
47 D. B. Redford, 'Foreigners (Especially Asiatics) in the Talatat' in D. B Redford, ed., *The Akhenaten Temple Project*, Vol. 2: *Rwd - mnw, Foreigners and Inscriptions*, Toronto (1988), 19; A. R. Schulman, 'Hittites, Helmets and Amarna', 54, 62, pl. 14; W. R. Johnson, *An Asiatic Battle Scene of Tutankhamun from Thebes*, 157–8 [15].
48 In the late 1950s and early 1960s, during restoration works of the first section of the Avenue of Sphinx that connects Luxor temple with Karnak Temple (M. Abdul-Qader, 'Preliminary Report of the Excavations carried out in the

temple of Luxor, Seasons 1958-1959 and 1959-1960', *ASAE* 60 (1968), 232-5; M. Boraik, 'Sphinxes Avenues Excavation: First Report', *CahKarn* 13 (2010), 46; *idem*, 'The Sphinxes Avenue Excavations to the East Bank of Luxor' in M. G. Folli, ed., *Sustainable Conservation and Urban Regeneration: The Luxor Example*, Cham (2018), 7-31), Mohammed Abdul-Qader Muhammad and Mahmud Abdul Razik have unearthed a large number of relief fragments from the Theban monuments attributed to Tutankhamon. Unfortunately, when these fragments were transported to the Luxor Temple, they were irremediably mixed up (W. R. Johnson and J. B. McClain, 'A Fragmentary Scene of Ptolemy XII Worshiping the Goddess Mut and Her Divine Entourage' in S. H. D'Auria, ed., *Servant of Mut: Studies in Honor of Richard A. Fazzini*, Leiden, Boston (2008), 134).

49 W. R. Johnson, *An Asiatic Battle Scene of Tutankhamun from Thebes*, 157-8 [15].
50 W. R. Johnson, *An Asiatic Battle Scene of Tutankhamun from Thebes*, 155-6 [8].
51 For Horemheb's usurpation and obliteration of Ay's monuments, see O. J. Schaden, 'Clearance of the tomb of Ay (WV-23)', *JARCE* 21 (1984), 39-64.
52 See U. Hölscher, *The Excavations of Medinet Habu*. Vol. II: *The Temples of the Eighteenth Dynasty*, Chicago (1939), 75-80; M. Gabolde, *Toutankhamon*, 428-30.
53 W. R. Johnson, *An Asiatic Battle Scene of Tutankhamun from Thebes*, 122-7.
54 The Epigraphic Survey, *The Temple of Khonsu*. Vol. I: *Scenes of King Herihor in the Court with Translation of Texts*, Chicago (1979), pl. 61.
55 W. R. Johnson, *An Asiatic Battle Scene of Tutankhamun from Thebes*, 125, 169-70 [60]; J. C. Darnell and C. Manassa, *Tutankhamun's Armies*, 182, fig. 24.
56 Although Yenoam has been depicted and mentioned several times in the Ramesside monuments, its precise location is debated. See N. Na'aman, 'Yeno'am', *Tel Aviv* 4 (1977), 168-77; M. G. Hasel, *Domination and Resistance: Egyptian Military Activity in the Southern Levant, ca. 1300-1185 BC*, Leiden (1988), 146-51; M. R. Abbas, 'The Town of Yenoam in the Ramesside War Scenes and Text of Karnak', *CahKarn* 16 (2017), 333, n. 10.
57 I. Rosellini, *I Monumenti Storici dell'Egitto e della Nubia. Tomo* II: *Monumenti Civili*, Pisa (1834), pl. 46; W. Wreszinski *Atlas* II, pl. 36; Y. Yadin, *Art of Warfare*, 230-1; A. J. Spalinger, 'The Northern War of Seti I: An Integrative Study', *JARCE* 16 (1979), 31-2; The Epigraphic Survey, *Relief and Inscriptions at Karnak*, Vol. IV: *The Battle Reliefs of King Sety* I, Chicago (1986), 35-6, pl. 23; J. R. Zorn, *BASOR* 360, 3-4, fig. 4; M. R. Abbas, *CahKarn* 16, 329-34, fig. 1-2.
58 I. Rosellini, *I Monumenti Storici: Tomo* I, pl. LIII; J.-F. Champollion, *Monuments de l'Égypte et de la Nubie d'après les dessins exécutés sur les lieux. Planches. Tome* III, Paris (1845), pl. CCXCV; W. Wreszinski *Atlas* II, pl. 53; A. J. Spalinger, *JARCE* 16, 41.
59 A profile of injuries suffered by the Asiatic warriors depicted in the battle scenes composing the decoration of the exterior face of the north wall of the Hypostyle

Hall at the Temple of Amun-Ra at Karnak, has been traced by Gonzalo M. Sanchez, a neurosurgeon with a long-life interest in Egyptology. See G. M. Sanchez, *JARCE* 37, 143–65.

60 I. Rosellini, *I Monumenti Storici: Tomo* I, pl. LXXXVII–XC; J.-F. Champollion, *Monuments de l'Égypte et de la Nubie d'après les dessins exécutés sur les lieux. Planches. Tome* I, Paris (1835), pl. 17, 23–5; J. H. Breasted, *The Battle of Kadesh: A Study in the Earliest Known Military Strategy*, Chicago (1903), 40–1, pl. VI; C. Kuentz, *La bataille de Qadech: les textes ('Poème de Pentaour' et 'Bulletin de Qadech') et les bas-reliefs*, Cairo (1928–34), 189–98, pl. XLII; W. Wreszinski *Atlas* II, pl. 170; Ch. Desroches-Noblecourt et al., *Le grand Temple d'Abou Simbel,* II. *La bataille de Qadech. Description et inscriptions. Dessins et photographies*, Cairo (1971), 29–31, pl. IV, XXX; C. Obsomer, *Ramsès II*, Paris (2012), 132, fig. 48.

61 This episode is described in the two primary accounts of the battle of Kadesh – the *Poem* (*KRI* II, 28–44) and the *Bulletin* (*KRI* II, 119, 6–123, 10) – and in the rhetorical texts that accompany the battle scenes in the Ramesseum (KRI II, 135, 1–136, 5) and the Temple of Abu Simbel (*KRI* II, 134, 9–14).

62 These iconographic details contradict the claim according to which 'the Qadeš reliefs do not provide evidence about Hittite helmets' (J. Lorenz and I. Schrakamp, 'Hittite Military Warfare' in H. Genz and D. P. Mielke, eds, *Insights into Hittite History and Archaeology*, Leuven, Paris, Walpole MA (2011), 141).

63 A. Wiese et al., *Antikenmuseum Basel und Sammlung Ludwig: Die Ägyptische Abteilung*, Mainz (2001), 102–3; L. Petersen et N. Kehrer, ed., *Ramses: Göttlicher Herrscher am Nil*, Karlsruhe (2016), 322 [179].

64 For an overview of the characteristics of this type of statuettes, see: G. Pinch, *Magic in Ancient Egypt*, London (1994), 90–102; E. Warmenbol, *Sphinx: les gardiens de l'Égypte*, Brussels (2006), 52, 193, 195; D. Ben-Tor, ed., *Pharaoh in Canaan: The Untold Story*, Jerusalem (2016), 144–5.

65 See J. B. Pritchard, 'Syrians as Pictured in the Paintings of the Theban Tombs', *BASOR* 122 (1951), 36–41; F. B. Anthony, *Foreigners in Ancient Egypt: Theban Tomb Paintings from the Early Eighteenth Dynasty*, London, New York (2017), 23.

66 S. Petschel and M. von Falck, eds, *Pharao siegt immer: Krieg und Frieden im Alten Ägypten*, Hamm (2004), 56–7.

67 A. M. Pollastrini, 'Una rappresentazione "tridimensionale" di tarda Età del Bronzo?' in A. Di Natale and C. Basile, eds, *Atti del XVIII Convegno di Egittologia e Papirologia, Siracusa, 20–23 Settembre 2018*, Siracuse (2020), 145–6.

68 See R. G. Morkot, 'War and the Economy: The International "Arms Trade" in the Late Bronze Age and after' in T. Schneider and K. Szpakowska, eds, *Egyptian Stories: A British Egyptological Tribute to Alan B. Lloyd on the Occasion of His Retirement*, Münster (2007), 173.

69 D. H. Lew, 'Manchurian Booty and International Law', *AJIL* 40.3 (1946), 586. See also W. G. Downey, Jr., 'Captured Enemy Property: Booty of War and Seized Enemy Property', *AJIL* 44.3 (1950), 488–9.
70 M. Liverani, *Guerra e Diplomazia nell'Antico Oriente. 1600–1100 a. C.*, Bari (1994), 208–9.
71 See below section 5.2 𓀀𓏥—𓊃 *mss n ꜥḥꜣ*.
72 For the most updated report of the battle of Megiddo, see the chapter 'Thutmose III: Strategic commander' in A. J. Spalinger, *Leadership Under Fire: The Pressure of Warfare in Ancient Egypt*, Paris (2022), 79–135.
73 G. A. Reisner, 'Inscribed monuments from Gebel Barkal', *ZÄS* 66 (1931), 80, (2); G. A. Reisner and M. B. Reisner, 'Inscribed monuments from Gebel Barkal, Part 2: The Granite Stela of Thutmose III', *ZÄS* 69 (1933), 32; J. A. Wilson, *ANET*, 238; J. K. Hoffmeier, 'The Gebel Barkal Stela of Thutmose III (2.2 B)' in W. W. Hallo and K. L. Younger, eds, *The Context of Scripture*, Vol. 2: *Monumental Inscriptions from the Biblical World*, Leiden, Boston, Köln (2000), 16.
74 See H. Grapow, *Studien zu den Annalen Thutmosis des Dritten und zu ihnen verwandten historischen Berichten des neues Reiches*, Berlin (1949), 11; A. J. Spalinger, 'A Critical Analysis of the "Annals" of Thutmose III (Stücke V–VI)', *JARCE* XIV (1977), 47; P. Grandet, *Les Pharaons*, 106.
75 É. Delange, *Monuments égyptiens du Nouvel Empire: La Chambre des Ancêtres, les Annales de Thoutmosis III et le décor de(s) palais de Séthi Ier*, Paris, 2015, 144–8.
76 See G. Legrain, 'La grande stèle d'Amenôthès II à Karnak', *ASAE* IV (1903), 126–31. Another version of the same text has been found reused as ceiling in a Twenty-second Dynasty tomb at Memphis. See A. M. Badawy, 'Die neue historische Stele Amenophis' II', *ASAE* XLII (1943), 1–23, pl. I; E. Edel, 'Die Stelen Amenophis' II aus Karnak und Memphis mit dem Bericht über den Libyerkrieg Merenptahs', *ZDPV* 69 (1953), 97–126.
77 J. H. Breasted, *Historical Records of Egypt* II, 305–9; J. A. Wilson, *ANET*, 245–7.
78 Sh. Yeivin, 'Amenophis II's Asianic Campaigns', *JARCE* 6 (1967), 121; P. Grandet, *Les Pharaons*, 116–17.
79 D. B. Redford, 'The Northern Wars of Thutmose III' in D. E. Cline and D. O'Connor, eds, *Thutmose III: A New Biography*, Ann Arbor (2006), 334.
80 See D. Panagiotopoulos, 'Foreigners in Egypt in the Time of Hatshepsut and Thutmose III' in D. E. Cline and D. O'Connor, eds, *Thutmose III: A New Biography*, Ann Arbor (2006), 372–3. See also E. Bleiberg, 'Commodity Exchange in the Annals of Thutmose III', *JSSEA* 11 (1981), 107–10; idem, 'The King's Privy Purse During the New Kingdom: An Examination of INW', *JARCE* 21 (1984), 155–67.
81 M. Liverani, *Guerra e Diplomazia*, 220–2.

82 The scenes of the Pharaoh seated in the royal kiosk were frequent in the Theban tombs of high-ranking officers during the Eighteenth Dynasty, especially under the reign of Thutmose IV and Amenhotep III. See M. K. Hartwig, *Tomb Painting and Identity in Ancient Thebes, 1419-1372 BCE*, Turnhout (2004), 55-73, 129-30, n. 66; B. M. Bryan, 'Pharaonic Painting through the New Kingdom' in A. B. Lloyd, ed., *A Companion to Ancient Egypt*, Chichester, UK (2010), 1002-3.

83 C. Aldred, 'The Foreign Gifts Offered to Pharaoh', *JEA* 56 (1970), 105; R. G. Morkot, 'War and the Economy: The International "Arms Trade" in the Late Bronze Age and After', 174.

84 Ph. Virey, *Sept Tombeaux thébains de la XVIIIe dynastie*, Paris (1891), 204, 206, fig. 1; N. De Garis Davies and N. De Garis Davies, *The Tomb of Menkherrasonb, Amenmose and Another (Nos. 86, 112, 42, 226)*, London (1933), 8 [63], 9 [80], pl. VII.

85 Ph. Virey, 'Le tombeau d'un Seigneur de Thini dans la nécropole de Thèbes', *RecTrav* 9 (1887), 27-32; idem, *Sept Tombeaux thébains*, 362-70; I. Morfini and M. Álvarez Sosa, 'The *Min Project*: First working seasons on the unpublished Tomb of Min (TT 109) and Tomb Kampp -327-: the Tomb of May and a replica of the Tomb of Osiris' in G. Rosati and M. C. Guidotti, eds, *Proceedings of the International Congress of Egyptologist XI, Florence, Italy 23-30 August 2015*, Oxford, 2017, 427-32.

86 N. De Garis Davies and N. De Garis Davies, *The Tomb of Menkherrasonb*, 29 [116], [136], pl. XXXIV, XXXV.

87 Ph. Virey, *Le tombeau de Rekhmara, Préfet de Thèbes sous la XVIIIe dynastie*, Paris (1889), 170-2; N. De Garis Davies, *The Tomb of Rekh-mi-Re' at Thebes*, Vol. I-II, New York (1943), 36-8 [19], pl. XXXVII.

88 N. De Garis Davies, *The Tomb of Ḳen-Amūn at Thebes*, Vol. I, New York (1930), 31 [114], pl. XXII.

89 Y. Yadin, *Art of Warfare*, 85; T. Dezsö, 'Scale Armour of the 2nd Millennium BC', 196-7.

90 The Myceanaean-era body armour is currently exhibited in the Nauplion Archaeological Museum (Inv-Nr 19.001-2). See A. M. Snodgrass, *Arms and Armour of the Greeks*, Ithaca, New York (1967), 24, fig. 9; P. Càssola Guida, *Le armi defensive dei Micenei nelle raffigurazioni*, Rome (1973), 52-3; pl. XV; H. W. Catling, 'Panzer', in H.-G. Buchholz and J. Wiesner, *Kriegwesen, Teil 1: Schutzwaffen und Wehrbauten, Archaeologia Homerica* E 1, Göttingen (1977), 96-102, pl. VII; E. Andrikou, 'New Evidence on Mycenaean Bronze Corselets from Thebes in Boeotia and the Bronze Age Sequence of Corselets in Greece and Europe' in I. Galanaki *et al.*, eds, *Between the Aegean and Baltic Seas: Prehistory across Borders: Proceedings of the International Conference Bronze and Early Iron Age*

Interconnections and Contemporary Developments between the Aegean and the Regions of the Balkan Peninsula, Central and Northern Europe, University of Zagreb, 11–14 April 2005, Liège, Austin (2007), 402, pl. Ca–b; M. Mödlinger, 'European Bronze Age Cuirasses: Aspects of Chronology, Typology, Manufacture and Usage', *JRGZ* 59 (2012), B. Molloy, 'The Origins of Plate Armour in the Aegean and Europe', *Talanta* 44 (2013), 274, 281–5, fig. 1.

91 T. D. Hulit, *Late Bronze Age Scale Armour in the Near East: An Experimental Investigation of Materials, Construction, and Effectiveness, with a Consideration of Socio-economic Implications*, PhD dissertation, Durham University (2002), 37–8, pl. 10.

92 L. Manniche, *Lost Tombs: A Study of Certain Eighteenth Dynasty Monuments in the Theban Necropolis*, London, New York (1988), 91–2.

93 *LD Tafelwerke Abt.* III, Band V, pl. 64.

94 T. D. Hulit, *Late Bronze Age Scale Armour*, 37.

95 Eg J. R. Zorn, *BASOR* 360, 6, fig. 5; W. Hovestreydt, 'Sideshow or not? On Side-Rooms of the First Two Corridors in the Tomb of Ramesses III' in B. J. J. Haring et al., eds, *The Workman's Progress: Studies in the Village of Deir el-Medina and other Documents from Western Thebes in Honour of Rob Demarée*, Leiden (2014), 119.

96 N. De Garis Davies, 'The Graphic Work of the Expedition at Thebes', *BMMA*, Vol. 23, No. 12, Part 2: The Egyptian Expedition 1927–8 (Dec. 1928), 44–9, fig. 6; F. Kampp, *Die thebanische Nekropole: zum Wandel des Grabgedankes von der XVIII. bis zur XX. Dynastie*, Teil I, Mainz (1996), 360–4; V. Angenot, 'Les peintures de la chapelle de Sennefer (TT 96A)', *Egypte* 45 (2007), 21–32.

97 See W. L. Moran, *The Amarna Letters*, Baltimore, London, (1992), 51; J. Aruz et al., ed., *Beyond Babylon: Art, Trade, and Diplomacy in the Second Millenium B. C.*, New Haven, London (2008), 159.

98 See below section 5.3 𓏏𓂋𓈖𓏤 *tryn*.

99 See F. Bisson de la Roque, *Rapport sur les Fouilles de Médamoud. Année 1929*, Cairo (1930), 43–4, fig. 36; R. Cottevieille-Giraudet, *Rapport sur les Fouilles de Médamoud: Les Reliefs d'Aménophi IV Akhenaton*, Cairo (1936), 19, 55, fig. 86; A. R. Schulman, 'Some Observations on the Military Background of the Amarna Period', *JARCE* 3 (1964), 53–4; W. R. Johnson, *An Asiatic Battle Scene of Tutankhamun from Thebes*, 155–6 [3].

100 See W. R. Johnson, *An Asiatic Battle Scene of Tutankhamun from Thebes*, 155–6 [4].

101 See G. Legrain, *Les temples de Karnak: fragments du dernier ouvrage de Georges Legrain*, Brussels (1929), 134–6, fig. 87; A. R. Schulman, *JARCE* 3 (1964), 55, n. 29; M. Eaton-Krauss, *MDAIK* 44, 5, n. 30; W. R. Johnson, *An Asiatic Battle*

Scene of Tutankhamun from Thebes, 165-6 [44]; J. M. Galán, 'Mutilation of Pharaoh's Enemy' in M. Eldamaty and M. Trad, eds, *Egyptian Museum Collections around the World: Studies for the Centennial of the Egyptian Museum, Cairo*, Vol. I, Cairo (2002), 442-3, fig. 1; J. Martínez Babón, *Historia Militar*, 96, fig. 5; A. M. Gnirs and A. Loprieno, 'Krieg und Literatur' in R. Gundlach and C. Vogel, eds, *Militärgeschichte des pharaonischen Ägypten, Altägypten und seine Nachbarkulturen im Spiegel der aktuellen Forschung*, Paderborn etc. (2009), 252; M. Gabolde, *Toutankhamon*, 417-18, fig. 185.

102 See W. R. Johnson, *An Asiatic Battle Scene of Tutankhamun from Thebes*, 12-13; M. Gabolde, *Toutankhamon*, 417-18, fig. 186. Another block found in the Second Pylon of the Temple of Karnak, depicting several Egyptian soldiers with bound Nubian captives, supports the hypothesis concerning the existence of a war scene commemorating a military expedition in Nubia (H. Chevrier, 'Rapport sur les travaux de Karnak 1952-1953', *ASAE* 53 (1953), 11, fig. 1, pl. I a-b).

103 G. T. Martin, *The Memphite Tomb of Ḥaremḥeb Commander-in-Chief of Tut ʿankhamūn*, Vol. I: *The Reliefs, Inscriptions and Commentary*, London, Turnhout (1989), 43-4, pl. 32, 34 [22].

104 As regards the military ranks held by Horemheb, see P.-M. Chevereau, *Prosopographie des Cadres Militaires Egyptiens du Nouvel Empire*, Paris (1994), 11-12, 15, 23-4, 29-30, 218.

105 Representations of horsemen are rare but not unknown in Egyptian art. See A. R. Schulman, 'Egyptian Representations of Horsemen and Riding in the New Kingdom', *JNES* 16 (1957), 263-71; M. A. Littauer and J. H. Crouwel, *Wheeled Vehicles and Ridden Animals in the Ancient Near East*, Leiden (1979), 96-7; C. Rommelaere, *Les chevaux du Nouvel Empire égyptien. Origines, races, harnacement*, Brussels (1991), 125, 128, fig. 94; R. Drews, *Early Riders: The Beginnings of Mounted Warfare in Asia and Europe*, New York, London (2004), 45-6.

106 See G. Vogelsang-Eastwood, *Pharaonic Egyptian Clothing*, Leiden, New York, Köln (1993), 24-31.

107 A. H. Gardiner, *Ancient Egyptian Onomastica*, Vol. I, Oxford (1947), 28*-29* [97].

108 A. H. Gardiner, *AEO*, Vol. I, 28* [96].

109 J. Yoyotte and J. López, 'L'organisation de l'armée et les titulatures des soldats au nouvel empire égyptien', *BiOr* 26.1-2 (1969), 10-11, basically followed by P.-M. Chevereau, *Prosopographie*, 173-90 and J. C. Darnell and C. Manassa, *Tutankhamun's Armies*, 64, n. 41.

110 A. R. Schulman, 'The Egyptian Chariotry: A Reexamination', *JARCE* 2 (1963), 88-9; idem, *Military Rank, Title and Organization in the Egyptian New Kingdom*, Berlin (1964), 67-8.

111 A. J. Spalinger, 'Egyptian Chariots: Departing for War' in A. J. Veldmeijer and S. Ikram, eds, *Chasing Chariots: Proceedings of the First Chariot Conference* (Cairo 2012), Leiden: Sidestone (2013), 238–9.

3 Personal Protective Equipment during the Ramesside Period

1 See R. Drews, *The End of the Bronze Age*, 135–63.
2 J.-F., Champollion, *Monuments de l'Égypte et de la Nubie: Tome* I, pl. XXIX; I. Rosellini, *I Monumenti Storici: Tomo* I, pl. C–CII; Ch. Desroches-Noblecourt et al., *Grand Temple d'Abou Simbel*, II, 9–14, pl. XIII, XV, XVIII–XIX.
3 As regards the role of Sherden warriors in the Ramesside royal guard, see M. R. Abbas, 'The Bodyguard of Ramesses II and the Battle of Kadesh', *ENiM* 9 (2016), 113–23; *idem*, 'A Survey of the Military Role of the Sherden Warriors in the Egyptian Army during the Ramesside Period', *ENiM* 10 (2017), 7–23.
4 For an overview of the equipment of the 'northern' warriors depicted on the Egyptian reliefs, see J. Vanschoonwinkel, 'Les Peuples de la Mer d'après une lecture archéologique des reliefs de Médinet Habou' in C. Karlshausen and C. Obsomer, eds, *De la Nubie à Qadech: La guerre dans l'Égypte ancienne*, Brussels (2016), 209–21.
5 W. Wreszinski, *Atlas* II, pl. 58, 58a; Y. Yadin, *Art of Warfare*, 228.
6 Frank Yurko first proposed that the Cour de la Cachette reliefs gave a pictorial representation of the victories over Ashkelon, Gezer, Yenoam and the Israelites celebrated in the Victory Stele of Merenptah (F. J. Yurko, 'Meremptah's Canaanite Campaign', *JARCE* 22 (1986), 189–215; *idem*, '3,200-Year-Old Picture of Israelites Found in Egypt', *BAR* 16.5 (September/October 1990), 20–38). Yurko's theory enjoys considerable popularity among scholars. See eg L. E. Stager, 'Merenptah, Israel and Sea Peoples: New Light on an Old Relief', *ErIsr* 18 (1985), 56–64; M. G. Hasel, *Domination and Resistance*, 49–50, 199–201; I. Singer, 'Merneptah's Campaign to Canaan & the Egyptian Occupation of the Southern Costal Plain of Palestine in the Ramesside Period', *BASOR* 269 (1988), 1–10; M. G. Hasel, 'Israel in the Merneptah Stela', *BASOR* 296 (1994), 45–61.
7 As regards the recurring depiction of Asiatics with raised arms and incense burners surrendering to the Pharaoh, see A. A. Burke, 'New Light on Old Reliefs: New Kingdom Egyptian Siege Tactics and Asiatic Resistance' in J. D. Schloen, ed., *Exploring the Longue Durée: Essays in Honor of Lawrence E. Stager*, Winona Lake, IN (2009), 61–3. It must be mentioned that Anthony Spalinger had

previously interpreted this pictorial *topos* as a form of Canaanite ritual meant to implore Baal's help. According to Spalinger, the aforementioned ritual included the infant sacrifice (A. Spalinger, 'A Canaanite Ritual Found in Egyptian Reliefs', *JSSEA* 8 (1978), 47–60). Moreover, William. W. Hallo later approved and expanded this latter interpretation (W. W. Hallo, 'A Ugaritic Cognate for Akkadian *hitpu?*' in R. Chazan, W. W. Hallo and L. H. Schiffman, eds, *Ki Baruch Hu: Ancient Near Eastern, Biblical, and Judaic Studies in Honour of Baruch A. Levine*, Winona Lake, IN (1999), 78–85).

8 A. H. Gardiner, *Late-Egyptian Miscellanies*, Brussels (1937), 27; R. A. Caminos, *Late-Egyptian Miscellanies*, London (1954), 96, 98. See also O. Herslund, 'Chariots in the Daily Life on New Kingdom Egypt: A Survey of Production, Distribution and Use in Texts' in A. J. Veldmeijer and S. Ikram, eds, *Chasing Chariots: Proceedings of the First International Chariot Conference* (Cairo 2012), Leiden (2013), 126; *idem*, 'VI. Chronicling Chariots: Texts, Writing and Language of New Kingdom Egypt' in A. J. Veldmeijer and S. Ikram, eds, *Chariots in Ancient Egypt: The Tano Chariot, a Case of Study*, Leiden (2018), 162.

9 𓊪𓊖𓉐𓏏𓅓𓈖𓇿𓂋𓅱 *P3 ḥtm n T3rw*, 'the Fortress of Tjaru' marked the border between Egypt proper and the hostile Asiatic lands. Tjaru served as a starting point for the Egyptian troops who travelled along the so-called 'Ways of Horus', the military route leading out of Egypt into Syro-Palestine. See J. K. Hoffmeier, 'Reconstructing Egypt's Eastern Frontier Defense Network in the New Kingdom (Late Bronze Age)' in F. Jesse and C. Vogel, eds, *The Power of Walls – Fortifications in Ancient Northeastern Africa: Proceeding of the International Workshop held at the University of Cologne 4th–7th August 2011*, Cologne (2013), 163–94.

10 A. M. Blackman and T. E. Peet, 'Papyrus Lansing: A Translation with Notes', *JEA* 11 (1925), 292; A. H. Gardiner, *Late-Egyptian Miscellanies*, 107–8; R. A. Caminos, *Late-Egyptian Miscellanies*, 401; M. Lichtheim, *Ancient Egyptian Literature*. Vol. II: *The New Kingdom*, Berkeley (1976), 172.

11 O. Herslund, 'VI. Chronicling Chariots: Texts, Writing and Language of New Kingdom Egypt', 126.

12 I. Rosellini, *I Monumenti Storici. Tomo I*, pl. CXXV; J.-F., Champollion, *Monuments de l'Égypte et de la Nubie. Tome III*, pl. CCXVIII; The Epigraphic Survey, *Medinet Habu*, Vol. I, pl. 29; G. A. Gaballa, *Narrative*, 122; W. J. Murnane, *The Road to Kadesh*, 13; J. Martínez Babón, *Historia Militar*, 52–3, fig. 19; M. R. Abbas, *ENiM* 10, 13.

13 W. F. Edgerton and J. A. Wilson, *Historical Records of Ramses III: The Texts in the Medinet Habou*, Volumes I and II, Chicago (1936), 35; A. J. Spalinger, *War in Ancient Egypt*, 253.

14 *KRI* V, 28, 14–16.

15 Concerning the alleged location of Djahy, see J. K. Hoffmeier, 'A Possible Location in Northwest Sinai for the Sea and Land Battles between the Sea Peoples and Ramesses III', *BASOR* 380 (2018), 1–25.
16 The Epigraphic Survey, *Medinet Habu*, Vol. I, pl. 32–4; W. Wreszinski, *Atlas*. II. *Teil*, pl. 121–2; G. A. Gaballa, *Narrative*, 122–3; N. K. Sandars, *The Sea Peoples: Warriors of the Ancient Mediterranean. 1250–1150 BC*, London (1985), 120–4; R. Drews, *The End of the Bronze Age*, 158, fig. 10; D. O'Connor, 'The Sea People and the Egyptian Sources' in E. D. Oren, ed., *The Sea Peoples and Their World: A Reassessment*, Philadelphia (2000), 95, fig. 5.5; J. Vanschoonwinkel, 'Les Peuples de la Mer d'après une lecture archéologique des reliefs de Médinet Habou', 197, fig. 1.
17 According to Silvio Curto, Ramesses III's military operations in the eastern Delta would be considered as 'la prima grande operazione antisbarco della storia' (S. Curto, L'arte militare presso gli antichi Egizi, Turin (1973), 10–12).
18 The Epigraphic Survey, *Medinet Habu*, Vol. I, pl. 36–40; W. Wreszinski, *Atlas*. II. *Teil*, pl. 115–17; H. H. Nelson, 'The Naval Battle Pictured at Medinet Habu', *JNES* 2.1 (1943), 40–55, fig. 4, pl. 1; G. A. Gaballa, *Narrative*, 122–3; N. K. Sandars, *The Sea Peoples*, 124–30, fig. 80–4; R. Drews, *The End of the Bronze Age*, 158–9, fig. 7; J. Vanschoonwinkel, 'Les Peuples de la Mer d'après une lecture archéologique des reliefs de Médinet Habou', 197–8, fig. 2.
19 The Epigraphic Survey, *Medinet Habu*, Vol. II, pl. 69–70; W. Wreszinski, *Atlas*. II. *Teil*, pl. 141–141a; P. Grandet, *Ramsés III: Histoire d'un regne*, Paris (1993), 209–12; C. Simon, 'Les campagnes militaires de Ramsès III à Médinet Habou: Entre vérité et propagande' in C. Karlshausen and C. Obsomer, eds, *De la Nubie à Qadech: La guerre dans l'Égypte ancienne*, Brussels (2016), 185–91.
20 The Epigraphic Survey, *Medinet Habu*, Vol. II, pl. 71–2; W. Wreszinski, *Atlas*. II. *Teil*, pl. 136–7; C. Simon, 'Les campagnes militaires de Ramsès III à Médinet Habou. Entre vérité et propagande', 185–91.
21 The Epigraphic Survey, *Medinet Habu*, Vol. II, pl. 90, W. Wreszinski, *Atlas*. II. *Teil*, pl. 153; C. Simon, 'Les campagnes militaires de Ramsès III à Médinet Habou. Entre vérité et propagande', 176.
22 The Epigraphic Survey, *Medinet Habu*, Vol. II, pl. 94; W. Wreszinski, *Atlas*. II. *Teil*, pl. 146–7; C. Simon, 'Les campagnes militaires de Ramsès III à Médinet Habou. Entre vérité et propagande', 176.
23 The Epigraphic Survey, *Medinet Habu*, Vol. II, pl. 88.
24 J.-F., Champollion, *Monuments de l'Égypte et de la Nubie: Tome* III, pl. CCXXVIII; The Epigraphic Survey, *Medinet Habu*, Vol. II, pl. 90; M. G. Hasel, *Domination and Resistance*, 51–2, fig. 7; C. Simon, 'Les campagnes militaires de Ramsès III à Médinet Habou. Entre vérité et propagande', 178–9, fig. 4. As regards the siege

techniques depicted in the relief, see A. M. Pollastrini, 'La poliorcetica in Egitto dall'Antico Regno alla XXV dinastia', 242–3; F. Monnier, *ENiM* 15, 72, fig. 11.

25 The exact location of the town is still debated by scholars. Michael Astour identifies Tunip with Hamat in Syria (M. C. Astour, 'The Partition of the Confederacy of Mukiš – Nuḫašše – Nii by Šuppiluliuma. A Study in Political Geography of the Amarna Age', *Orientalia* 39 (1969), 394–5; *idem*, 'Tunip-Hamat and Its Region', *Orientalia* 46 (1977), 57 ff.). Alternatively, Horst Klengel identified the town with Tell Asharinah on the middle course of the river Orontes (H. Klengel, 'Tunip und andere Probleme der historischen Geographie Mittelsyriens' in K. van Lerberghe and A. Schoors, eds, *Immigration and Emigration within the Ancient Near East: Festschrift E. Lipinski*, Leuven (1995), 128). See also W. H. van Soldt, 'The Orontes Valley in texts from Alalaḫ and Ugarit during the Late Bronze Age, ca 1500–1200 BC', *Syria* IV (2016), 141–2.

26 B. E. J. Peterson, *Zeichnungen aus einer Totenstadt: Bildostraka aus Theben-West, ihre Fundplätze, Themata und Zweckbereiche litsamt einem Katalog der Gayer-Anderson-Sammlung in Stockholm*, Stockholm (1973), 93, [81], pl. 45; A. Eggebrecht, ed., *Ägyptens Aufstieg zur Weltmacht*, Mainz (1987), 122, n. 22; S. Petschel and M. von Falck, eds, *Pharao siegt immer*, 104, n. 94; A. Herold, 'Aspekte ägyptischer Waffentechnologie – von der Frühzeit bis zum Ende des Neuen Reiches' in R. Gundlach and C. Vogel, eds, *Militärgeschichte des pharaonischen Ägypten: Altägypten und seine Nachbarkulturen im Spiegel der aktuellen Forschung*, Padeborn (2009), 231, n. 189.

27 For an overview of the tomb of Ramesses III, see F. Mauric-Barberio, 'La tombe de Ramsès III', *Egypte* 34 (2004), 15–34.

28 Starting from the French campaign in Egypt (1798–1801), the decoration of the tomb of Ramesses III caught the attention of scholars (F. Mauric-Barberio, 'Reconstitution du décor de la tombe de Ramsès III (partie inférieure) d'après les manuscrits de Robert Hay', *BIFAO* 104 (2004), 389, note 2). So it is little wonder that the helmets and the weapons painted in Room M were copied and reproduced several times in nineteenth-century Egyptological books. See e.g. *Description de l'Égypte: Antiquités, Planches. Tome* II, pl. 88; J.-F. Champollion, *Monuments de l'Égypte et de la Nubie, Tome* III, pl. CCLXII–CCLXIII; I. Rosellini, *I Monumenti Civili: Tomo* II, pl. CXXI; J. G. Wilkinson, *The Manner and Customs of the Ancient Egyptians*, Vol. I, London (1878), 219, fig. 52, 221, fig. 53a; E. Lefébure, *Les hypogées royaux de Thèbes, seconde division: Notices des hypogées*, Paris (1889), 96–7.

29 See W. Hovestreydt, 'Sideshow or not? On Side-Rooms of the First Two Corridors in the Tomb of Ramesses III', 118–19.

30 (K_1) KRI II, 28, 7–29, 1; J. H. Breasted, *Ancient Records of Egypt*, Vol. III, *The Nineteenth Dynasty*, Chicago (1906), 104, §312; E. de Rougé, *Oeuvres diverses*, Tome V, Paris (1914), 312; S. Hassan, *Le poème dit de Pentaour et le rapport officiel sur la bataille de Qadesh*, Cairo (1929), pl. 36, A, K, I; C. Kuentz, *La bataille*, 29, 237–8, §77–78; R. O. Faulkner, 'The Battle of Kadesh', *MDAIK* 16 (1958), 103.

31 (K_2) KRI II, 28, 8; C. Kuentz, *Bataille*, 56, 237–8, §§77–78.

32 (L_1) KRI II, 28, 9–29, 2; J. de Rougé, *Inscriptions hiéroglyphiques copiées en Égypte pendant la mission scientifique de M. le Vicomte Emmanuel de Rougé*, Vol. IV, Paris (1879), pl. CCXXXVI; S. Hassan, *Poème*, pl. 36 A, L I. Col. 22; J. H. Breasted, *Ancient Records*, Vol. III, 104, §312; C. Kuentz, *Bataille*, 76, 237–8, §§77–78; R. O. Faulkner, *MDAIK* 16 (1958), 103; A. H. Gardiner, *The Qadesh Inscriptions of Ramesses II*, Oxford (1960), 9.

33 KRI II, (L_2) 28, 10–29, 3; E. de Rougé, *Oeuvres diverses*, 312; S. Hassan, *Poème*, pl. 36 A, L III, Col. 22; J. H. Breasted, *Ancient Records*, Vol. III, 104, §312; C. Kuentz, *Bataille*, 121, 237–8, §§77–78; R. O. Faulkner, *MDAIK* 16 (1958), 103; A. H. Gardiner, *Qadesh*, 9. We must also add S. Hassan, *Poème*, pl. 36 A, L III, Col. 22, although the author confused the copy of the *Poem* (L_2) engraved on the exterior wall of the Court of Ramesses II with the copy of the *Poem* (L_3) engraved on the exterior wall of the Court of Amenhotep III (PM II², 334, [207]). Unfortunately, that specific part of *Poem* (L_3) has long since desappeared (G. Loukianoff, 'Un troisième texte du Poème de Pentaour sur la face oust du Temple de Louxor' *BIE* 9 (1926), 57–9, fig. 2).

34 KRI II, (R) 28.11; C. Kuentz, *Bataille*, 170, 237, §77.

35 KRI II, (Ch B_1) 28.11–29.5, (Ch B_2) 28.13–29.6; A. H. Gardiner, *Hieratic Papyri in British Museum*, Vol. I–II, 3rd Series, London (1935), 23–4, pl. 9–9_a, 10–10_a.

36 KRI II, (Rf – S) 28.1–29.4; E. de Rougé, 'Le Poëme de Pentaour, nouvelle tradution', *RecTrav* I (1870), 1–9 + facsimile; E. A. W. Budge, *Facsimiles of Egyptian Hieratic Papyri in the British Museum, 2nd Series* London (1923), 32–3, pl. LXXVII; C. Kuentz, *Bataille*, 237, §77.

37 Here, as well as at another point in the papyrus *Sallier III* (9,2), the term 𓎛𓎛𓎛 *ḥkr.w*, which basically means 'ornaments' (*Wb* III, 401, 15), is oddly replaced by a corrupted variant of the term See C. Kuentz, *Bataille*, 206; A. Spalinger, *The Transformation of an Ancient Egyptian Narrative: P. Sallier III and the Battle of Kadesh*, Wiesbaden (2002), 16.

38 KRI II, (L_1) 119, 12–120, 2; H. Brugsch, *Recueil de monuments égyptiens dessinés sur lieux, 1ere partie*, Cairo (1862), pl. XLI, col.23; J. H. Breasted, *Ancient Records*, Vol. III, 147, §326; C. Kuentz, *Bataille*, 100, 354, §86; R. O. Faulkner, *MDAIK* 16 (1958), 101; A. H. Gardiner, *Qadesh*, 30.

39 For the meaning of the word 🐝👁 ꜣ.t (Wb I, 1–2) see A. H. Gardiner, 'The First Two Pages of the "Wörterbuch"', JEA 34 (1948), 13–15.
40 KRI II, (L₂) 119, 13–120, 3; J. H. Breasted, *Ancient Records*, Vol. III, 147, §326; C. Kuentz, *Bataille*, 142, 354, §86; R. O. Faulkner, MDAIK 16 (1958), 101; A. H. Gardiner, *Qadesh*, 30.
41 KRI II, (R₁) 119, 14–120, 4; LD III, pl. 153; C. Kuentz, *Bataille*, 354, §86; A. H. Gardiner, *Qadesh*, 30.
42 KRI II, (I) 119, 15–120, 5; LD III, pl. 187; C. Kuentz, *Bataille*, 354, §86; A. H. Gardiner, *Qadesh*, 30; Ch. Desroches-Noblecourt et al., *Le grand Temple d'Abou Simbel*, II, 48–9.
43 H. te Velde, *Seth, God of Confusion: A Study of His Role in Egyptian Mythology and Religion*, Leiden (1967), 109; see also J. Zandee, 'Seth als Sturmgott', ZÄS 90 (1963), 148–9; N. Allon, 'Seth is Baal – Evidence from the Egyptian Script', ÄgLev 17 (2007), 15–22.
44 For the general bibliography concerning the so-called 'Year 400 Stela', refer to I. Cornelius, *The Iconography of the Canaanite Gods Reshef and Ba'al: Late Bronze and Iron Age I Periods (c 1500–1000 BCE)*, Freiburg, Switzerland–Göttingen, Germany (1994), 147–8, pl. 35.
45 See A. Dodson and D. Hilton, *The Complete Royal Families of Ancient Egypt*, Cairo (2004), 162.
46 See N. Allon, ÄgLev 17, 19; V. Lacroix, 'Le "Seth asiatique" de Ramsès II, origine et justification d'un culte' in *4e édition du colloque étudiant Jean-Marie Fecteau. 27, 28, 29 mars 2019, Université du Québec à Montréal, Actes* – Vol. 2, Revue Histoire, Idées, Sociétés, Open access, https://revuehis.uqam.ca/colloquejeanmariefecteau/le-seth-asiatique-de-ramses-ii-origine-et-justification-dun-culte/.
47 The myth of the Storm-god's fight against the deified Sea is a well-known *topos* in the Ancient Near East literature. See A. H. Sayce, 'The Astarte Papyrus and the Legend of the Sea', JEA 19.1/2 (1933), 56–9; Ph. Collombert and L. Coulon, 'Les dieux contre la mer. Le début du "papyrus d'Astarté" (pBN 202)', BIFAO 100 (2000), 219–21; A. R. W. Green, *Storm-God in the Ancient Near East*, Winona Lake, IN (2003), 178–88; N. Ayali-Darshan, *Storm-God and the Sea: The Origin, Version, and Diffusion of a Myth throughout the Ancient Near East*, Tübingen (2020), 1–10.
48 For the general bibliography concerning the *Astarte Papyrus*, see Ph. Collombert and L. Coulon, BIFAO 100, 193, n. 2.
49 T. Schneider, 'Foreign Egypt: Egyptology and the Concept of Cultural Appropriation', ÄgLev 13 (2003), 161.
50 See C. Kuentz, *Bataille*, 173–80; pl. XLI.

51 A. Prisse d'Avennes, *Histoire de l'art égyptien: d'après les monuments; depuis les temps les plus reculés jusqu'à la domination romaine*, Atlas, Tome II, Paris (1878), pl. III. 30.
52 See W. Wreszinski, *Atlas. II. Teil*, pl. 107–9; A. A.-H. Youssef *et al.*, *Ramesséum IV. Les batailles de Tounip et de Dapour*, Cairo (1977), pl. X–XI; A. Spalinger, 'Re-Reading Egyptian Military Reliefs' in M. Collier and S. Snape, eds, *Ramesside Studies in Honour of K. A. Kitchen*, Bolton (2011), 491, fig. 15; C. Obsomer, *Ramsès II*, 177–8, fig. 51.
53 W. Wreszinski, *Atlas. II. Teil*, pl. 78; C. Obsomer, *Ramsès II*, 178–81, fig. 52(a).
54 It is not clear how many times Ramesses II besieged Dapur or when. According to Kenneth A. Kitchen, Ramesses II took the town twice, during two separate Asiatic campaigns, in Years 8 and 10 of his reign (K. A. Kitchen, *Pharaoh Trimphant: The Life and Times of Ramesses II*, Warminster (1982), 68–70). More recently, Pierre Grandet and Claude Obsomer pointed out difficulties in determining the exact location of Dapur and whether the town was been besieged in both military campaigns (P. Grandet, *Pharaons*, 239–44; C. Obsomer, *Ramsès II*, 187–9).
55 *KRI* II, 175 K. Sethe, 'Mißverstandene Inschriften. 2. Ramses II. im Gebiete von Tunip', *ZÄS* 44 (1907), 36–9.
56 A. J. Spalinger, *War in Ancient Egypt*, 226.
57 C. Obsomer, *Ramsès II*, 177–9.
58 J.-F. Champollion, *Monuments de l'Égypte et de la Nubie*, Tome III, pl. CCLXII–CCLXIII; W. Hovestreydt, 'Sideshow or not? On Side-Rooms of the First Two Corridors in the Tomb of Ramesses III', 118–19.
59 We must not forget the mythological combat between Seth and Apophis. See G. Nagel, 'Set dans la barque solaire', *BIFAO* 28 (1929), 33–9; J. Zandee, *ZÄS* 90, 155–6; H. te Velde, *Seth, God of Confusion*, 99–108, pl. VII–VIII.

4 The Manufacture of Protective Gear in New Kingdom Egypt

1 We here allude to the archive found during the excavations of the Mesopotamian town of Nuzi (modern Yorghan Tepe, Iraq). Among the thousands of unearthed clay tablets, dating roughly from the mid-fifteenth to the mid-fourteenth century BCE, at least eighty include Akkadian terms relating to sixteen different types of helmets, corslets and their components. For an essential bibliography, see below Chapter 5 '.Terminology Relating to Protective Gear', note n. 24.
2 J. E. Quibell, *Excavations at Saqqara (1908-9, 1909-10). The Monastery of Apa Jeremias*, Cairo (1912), 9, pl. 68, 75; S. Sauneron, 'La manufacture d'armes de

Memphis', *BIFAO* 54 (1954), 10–11, fig. 2; G. T. Martin, *The Hidden Tomb of Memphis*, London (1991), 201–3; B. McDermott, *Warfare in Ancient Egypt*, Gloucestershire (2004), 142–3, fig. 90–91; N. Staring, *The Saqqara Necropolis through the New Kingdom: Biography of an Ancient Egyptian Cultural Landscape*, Leiden (2022), 376.

3 For an overall discussion and description of the restored decorative programme of Ky-iry's tomb, see W. Grajetzki, 'Das Grab des Kii–iri in Saqqara', *JEOL* 37, 2001–2 (2003), 114–16. See also A. Herold, *Streitwagentechnologie in der Ramses-Stadt: Knäufe. Knöpfe und Scheiben aus Stain*, Mainz (2006), 72–3, and L. Sabbahy, 'V. Moving Pictures: Context of Use and Iconography of Chariots in the New Kingdom' in A. J. Veldmeijer and S. Ikram, eds, *Chariots in Ancient Egypt: The Tano Chariot, a Case Study*, Leiden (2018), 148, fig. VI.4.

4 J. E. Quibell, *Excavations*, pl. 68, 75, 78; R. Drenkhahn, *Die Handwerker und ihre Tätigkeiten im Alten Ägypten*, Wiesbaden (1976), 131–2; A. Herold, 'Ein Puzzle mit zehn Teilen – Waffenkammer und Werkstatt aus dem Grab des *Ky-jrj* in Saqqara' in N. Kloth et al., eds, *Es werde niedergelegt als Schriftstück: Festschrift für Hartwig Altenmüller zum 65. Geburtstag*, Hamburg (2003), 196–7.

5 See A. H. Gardiner, *Egyptian Hieratic Texts: Transcribed, Translated and Annoted. Series I: Literary Texts of the New Kingdom. Part 1: The Papyrus Anastasi I and the Papyrus Koller, together with the Parallel Texts*, Leipzig (1911) 28*, 37, 11–13; W. Fischer-Elfert, *Die Satirische Streitschrift des Papyrus Anastasi I. Textzusammenstellung*, Wiesbaden (1983), 130; idem, *Die Satirische Streitschrift des Papyrus Anastasi I. Übersetzung und Kommentar*, Wiesbaden (1986), 227; E. Bresciani, *Letteratura e poesia dell'antico Egitto*, 4th edn, Turin (2007), 353.

6 In the decorative programme of New Kingdom private tombs, it is not unusual to find foreign offering bearers pulling chariots or carrying them on their shoulders. The same cannot be said about Egyptian bearers. Apart from the scene carved in Ky-iry's tomb, only an Eighteenth Dynasty relief fragment, kept in Yale Art Gallery (YAG ILE 1998.5.2), shows several Egyptian bearers pulling chariots. See C. Manassa, 'Two Unpublished Memphite Relief Fragments in the Yale Art Gallery', *SAK* 30 (2002), 257–8.

7 Metalworking has been quite a popular theme in private tombs since the Ancient Kingdom. Wall paintings, reliefs and hieroglyphic inscriptions provide scholars with a large amount of data concerning the steps of metal manufacturing and production. The major studies are J.-L. Chappaz, 'La purification de l'or', *BSEG* 4 (1980), 19–25; L. Garenne-Marot, 'Le travail du cuivre dans l'Égypte pharaonique d'après les peintures et les bas-reliefs', *Paléorient*, 11.1 (1985), 85–100; B. Scheel, 'Studien zum Metallhandwerk im Alten Ägypten I: Handlungen und Beischriften in den Bildprogrammen der Gräber des Alten Reiches', *SAK* 12 (1985), 117–77; *ead.*, 'Studien zum Metallhandwerk im Alten Ägypten II:

Handlungen und Beischriften in den Bildprogrammen der Gräber des Mittleren Reiches', *SAK* 13 (1986), 181–205; *ead.*, 'Studien zum Metallhandwerk im Alten Ägypten III: Handlungen und Beischriften in den Bildprogrammen der Gräber des Neuen Reiches und der Spätzeit', *SAK* 14 (1987), 247–64; *ead.*, *Egyptian Metalworking and Tools*, Buckinghamshire (1989); T. G. H. James, 'Gold Technology in Ancient Egypt: Mastery of Metal Working Methods', *Gold Bulletin* 5.2 (1972), 38–42; C. J. Davey, 'Old Kingdom Metallurgy in Memphite Tomb Images' in L. Evans, ed., Ancient Memphis: *'Enduring is the Perfection'*: *Proceedings of the International Conference held at Macquarie University, Sydney on August 14–15, 2008*, Leuven (2012), 85–107; P. Marini, 'Una scena di metallurgia e oreficeria dalla tomba M.I.D.A.N.05 a Dra Abu El-Naga', *EVO* 37 (2014), 89–100.

8   See below Chapter 5. 'Terminology Relating to Protective Gear'.

9   For example: a relief fragment from an unidentified New Kingdom tomb showing an arrow-maker checking the straightness of an arrow (G. T. Martin, *Corpus of Reliefs of the New Kingdom from the Memphite Necropolis and Lower Egypt*, Vol. I, London (1987), 18, n. 32 and pl. 10, n. 32; R. B. Partridge, *Fighting Pharaohs: Weapons and Warfare in Ancient Egypt*, Manchester (2002), 43, fig. 60; M. Gabolde, 'Toutânkhamon et les roseaux de Djapour' in C. Zivie-Coche and I. Guermeur, eds, *'Parcourir l'éternité'*: *Hommages à Jean Yoyotte*, Turnhout (2012), 457; *ead.*, *Toutankhamon*, Paris (2015), 330, fig. 144); a relief fragment from the late Eighteenth Dynasty tomb of the Overseer of Craftsmen, Chief of the Goldsmiths of the King, Ipuia, Teti Pyramid Cemetery at Saqqara (PM III^2/2, 55), now in the Egyptian Museum in Cairo (RT 17/6/24/12), showing artisans making statues, chariots, stelae and other objects (J. E. Quibell and A. G. K. Hayter, *Excavations at Saqqara, Teti Pyramid, North Side*, Cairo (1927), 10–11, 32, pl. 13; S. Sauneron, *BIFAO* 54 (1954), 9–10; S. Curto, *L'arte militare presso gli antichi Egizi*, 21; A. Herold, 'Ein Puzzle mit zehn Teilen – Waffenkammer und Werkstatt aus dem Grab des *Ky-jrj* in Saqqara', 199–200, fig. 3; N. Staring, *The Saqqara Necropolis*, 221, 223, 381); a relief fragment from a Memphite tomb possibly dating back to the Saite Period, now in the Museo Egizio of Florence (Inv. N. 2606), showing craftsmen engaged in a wide range of artisanal activities (M. C. Guidotti and F. Pecchioli Daddi, *La battaglia di Qadesh: Ramesse II contro gli Ittiti per la conquista della Siria*, Livorno (2002), 57; Y. Volokine, *La Frontalité dans l'iconographie de l'Égypte ancienne*, Geneva (2000), 35, pl. 39; G. Andreu-Lanoë *et al.*, eds, *L'art du contour: Le dessin dans l'Égypte ancienne*, Paris (2013), 120).

10 In Memphite necropoleis, several tomb owners hold titles relating to chariotry, suggesting that a chariot unit was accommodated in Memphis. See J. Malek, 'An Early Eighteenth Dynasty Monument of Sipair from Saqqâra', *JEA* 75 (1989), 73–4.

11 See Y. Yadin, *Art of Warfare*, 86-7; J. K. Hoffmeier, 'Observations on the Evolving Chariot Wheel in the 18th Dynasty', *JARCE* 13 (1976), 43-5; M. A. Littauer and J. Crouwel, 'An Egyptian Wheel in Brooklyn', *JEA* 65 (1979), 110-20; C. Manassa, *SAK* 30 (2002), 260-1.

12 O. Herslund, 'Chariots in the Daily Life of New Kingdom Egypt: A Survey of Production, Distribution and Use in Texts', 124; *idem*, 'VI. Chronicling Chariots: Texts, Writing and Language of New Kingdom Egypt', 161.

13 Nine copies of the *Onomasticon of Amenemope* are known. The title *irw ṯryn* appears in three copies, namely in the *Golenischeff Onomasticon* = P. Moscow 169 (A. H. Gardiner, *AEO*, Vol. I, 68*, 164, pl. IX A-IX), in the *Papyrus Hood* = P. BM EA 10202 (G. Maspero, 'Un manuel de hiérarchie égyptienne' *JournAs* 8ème série 11 (1888), 260; A. H. Gardiner, *AEO*, Vol. I, 68*, 164) and in a papyrus fragment found in the Ramesseum (W. Spiegelberg, *Hieratic Ostraka and Papyri found by J. E. Quibell in the Ramesseum, 1895-6*, London (1898), XLVII k, XLVIIa k; A. H. Gardiner, *AEO*, Vol. I, 68*, 164, pl. XX).

14 A. H. Gardiner, *AEO*, Vol. I, 68*-69*.

15 PM VII, 137; D. Randall-Maciver and C. L. Woolley, *The Eckley B. Coxe Junior Expedition to Nubia*: Vol. VII, *Buhen (text)*, Philadelphia (1911), 81; H. S. Smith, *The Fortress of Buhen of Buhen: The Inscriptions*, London (1976), 213, pl. LXXXI, 4.

16 For the Metropolitan Museum excavations at Malqata, see H. E. Winlock, 'The Work of the Egyptian Expedition', *BMMA* 7 (1912), 184-90; H. G. Evelyn-White, 'The Egyptian Expedition 1914-15', *BMMA* 10 (1915), 253-6; A. Lansing, 'Excavation at the Palace of Amenhotep III at Thebes', *BMMA* 13 Supplement (1918), 8-14; H. E. Winlock, 'Excavations at Thebes, 1919-20', *BMMA* 15 Supplement (1920), 12-32; W. C. Hayes, *The Scepter of Egypt*. Vol. II. *The Hyksos Period and the New Kingdom (1675-1080 BC)*, Cambridge, MA (1959), 244-55.

17 W. C. Hayes, *Scepter*, 254-5; C. A. Keller, 'Problems in Dating Glass Industries of the Egyptian New Kingdom: Examples from Malkata and Lisht', *JGS* 25 (1983), 19-23; A. K. Hodgkinson, 'High-Status Industries in the Capital and Royal Cities of the New Kingdom' in A. Hudecz and M. Petrik, eds, *Commerce and Economy in Ancient Egypt: Proceedings of the Third International Congress for Young Egyptologists 25-27 September 2009, Budapest*, Oxford (2010), 71-2; *idem*, *Technology and Urbanism in Late Bronze Age Egypt*, Oxford (2012), 214-15.

18 The scales from Malqata are mentioned for the first time in W. M. F. Petrie, *Tools and Weapons illustrated by the Egyptian Collection in University College, London, and 2,000 Outlines from Other Sources*, London (1917), 38, pl. XLII, n. 105-8. Although this reference is reported also in T. Dezsö, *RLA*, Band 10, 320, the scales shown in the pl. XLII of *Tools and Weapons*, quite strangely, do not match the genuine bronze plates stored in the MMA of New York. However, the scales from

Malqata have been extensively published over the years. See eg, Y. Yadin, *Art of Warfare*, 197; H. R. Robinson, *Oriental Armour*, London (1967), 4, fig. 2; T. D. Hulit, *Late Bronze Age Scale Armour*, 145; T. Dezsö, *RLA, Band* 10, 320; F. De Backer, *ResAnt* 8, 83, fig. 43; C. Trimm, *Fighting*, 543–4, fig. 7.23.

19 H. E. Winlock, *The Rise and Fall of the Middle Kingdom in Thebes*, New York (1947), 162–3, pl. 29. According to Winlock, the MMA kept several other armour scales from other Egyptian sites. Only scales MMA 22.1.46 (of which I have been unable to obtain information) and MMA 34.1.73 *a–c* are still part of the Museum collection. The other scales (MMA 09.183.3; 11.151.183–5, 192), according to Julie Zeftel, Senior Manager of Rights and Permissions at MMA, have been de-accessioned from the Museum's collection (personal communication, 12 November 2014). Due to a lack of further information, we are unable to say more about these 'lost' bronze scales.

20 As regards the ancient metal punching process, see W. Rostoker, 'Ancient Techniques for Making Holes in Sheet Metal', *AJA* 90 (1986), 93–4.

21 See Y. Yadin, *Art of Warfare*, 197.

22 In the context of the Asiatic tradition of armour-making, the practice of covering metallic helmets and body armour in velvet, brocade or silk was not uncommon. The aim of this practice must be sought in an attempt to make protective clothing more comfortable and less prone to overheating. Moreover, the use of fine cloth embodied the desire to show off wealth and luxury. See H. R. Robinson, *Oriental Armour*, 27, 34, 64, 79, 148–51.

23 R. F. S. Starr, *Nuzi: Report on the Excavations at Yorgan Tepa near Kirkuk, Iraq*, Vol. I, Cambridge, MA (1939), 476–7, pl. 126 A–B.

24 T. D. Hulit, *Late Bronze Age Scale Armour*, 78.

25 I embraced the classification system given by Timothy Kendall in his doctoral dissertation *Warfare and Military Matters in the Nuzi Tablets*, Brandeis University (1974). Kendall has categorized the armour scales from Nuzi into seven types according to their shapes and sizes.

26 I employed here the measurement provided by Richard Francis Strong Starr (R. F. S. Starr, *Nuzi*, Vol. I, 477). However, it must be said that, according to Thomas Hulit, Type 2 scales measure approximately 10.8 centimetres.

27 T. Kendall, *Warfare and Military Matters*, 274–6.

28 R. F. S. Starr, *Nuzi*, Vol. I, 126; T. Kendall, *Warfare and Military Matters*, 276.

29 T. D. Hulit, *Late Bronze Age Scale Armour*, 82.

30 W. Ventzke, 'Zur Rekonstruktion eines bronzenen Schuppenpanzer' in R. Hachmann, ed *Frühe Phöniker im Libanon: 20 Jahre deutsche Ausgrabungen in Kāmid el-Loz*, Mainz (1983), 96–7, fig. 45; R. Miron, *Kāmid el-Lōz. 10. Das 'Schatzhaus' im Palastbereich: Die Funde*, Bonn (1990), 65–72, pl. 15.

31 W. Ventzke, 'Zur Rekonstruktion eines bronzenen Schuppenpanzer', 97, fig. 49–50.
32 See W. C. Hayes, 'The Egyptian Expedition: The Excavation at Lisht', *BMMA* Vol. 29, No. 11, Part 2: *The Egyptian Expedition 1933–1934*, (Nov. 1934), 8, fig. 12–13; D. Arnold, *The Pyramid of Senwosret I*, New York (1988), 99–105; T. D. Hulit, *Late Bronze Age Scale Armour*, 145.
33 See J. von Beckerath, *Handbuch der ägyptischen Königsnamen*, Mainz (1988), 144–5.
34 See B. Thordeman, *Armour of the Battle of Wisby*, Uppsala (1939), 277, n. 255, fig. 232 [60–1]; W. C. Hayes, *Scepter*, 189–90; D. Arnold, *The Pyramid*, 104, fig. 49 [11–12], pl. 71, 72 *f–g*.
35 We adopted here the accession numbers reported by Dieter Arnold (D. Arnold, *The Pyramid*, 104). It should, however, be noted that the scales from Lisht are provided with different accession numbers (MMA 34.1.73 *a–c*) on the MMA web page www.metmuseum.org/art/collection/search/555562 (Accessed: 6 May 2023).
36 See M. Bietak, 'Harbours and Coastal Military Bases in Egypt in the Second Millennium BC: Avaris, Peru-nefer, Pi-Ramesse' in H. Willems and J.-M. Dahms, eds, *The Nile: Natural and Cultural Landscape in Egypt*, Bielefeld (2017), 62–3.
37 Facsimile of the papyrus *Anastasi* III: A. H. Gardiner, *Late-Egyptian Miscellanies*, XIV, 28.
38 The site Q IV of Pi-Ramesses/Qantir has been extensively discussed in a series of preliminary reports. For a general overview, see in particular E. B. Pusch, 'Vorbericht über die Abschlusskampagne am Grabungsplatz Q IV 1997', *ÄgLev* 9 (1999), 17–37; A. Herold, 'Piramesses – The Northern Capital: Chariots, Horses and Foreign Gods', in J. Goodnick Westenholz, ed., *Capital Cities: Urban Planning and Spiritual Dimensions. Proceedings of the Symposium held on May 27–29, 1996, Jerusalem, Israel*, Jerusalem (1998), 129–46; idem, *Streitwagentechnologie in der Ramses-Stadt: Bronze an Pferd und Wagen*, Mainz (1999); idem, *Knäufe, Knöpfe und Scheiben aus Stein*.
39 See E. B. Pusch, 'Metallverarbeitende Werkstätten der frühen Ramessidenzeit in Qantir-Piramesse/Nord', *ÄgLev* 1 (1990), 75–113; idem, 'Divergierende Verfahren der Metallverarbeitung in Theben und Qantir? Bemerkungen zu Konstruktion und Technik', *ÄgLev* 4 (1994), 145–70.

As regards the production and casting of bronze in Pi-Ramesses, see E. B. Pusch, 'High Temperature Industries in the Late Bronze Age Capital Piramesse (Qantir): II. A quasi-industrial bronze factory installation, tools and artifacts' in F. A. Esmael, ed., Z. A. Hawass, dir, *Proceedings of The First International Conference on Ancient Egyptian Mining & Metallurgy and Conservation of Metallic Artifacts, Cairo, Egypt, 10–12 April 1995*, Cairo (1996), 121–32; F. W. Rademakers *et al.*, 'Bronze Production in Pi-Ramesse: Alloying Technology and Material Use' in E. Ben-Yosef, ed., *Mining for Ancient Copper. Essays in Memory of Beno Rothenberg*, University Park, PA/Tel Aviv (2018), 503–25.

40 See A. Herold, 'High Temperature Industries in the Late Bronze Age Capital Piramesse (Qantir): III. Workshops of The Temple of Amun Qantir/Piramesse, Site Q-I, Stratum B/2' in F. A. Esmael, ed., Z. A. Hawass, dir, *Proceedings of The First International Conference on Ancient Egyptian Mining & Metallurgy and Conservation of Metallic Artifacts, Cairo, Egypt, 10–12 April 1995*, Cairo (1996), 133–45; S. Prell, 'A Glimpse into the Workshops of the Chariotry of Qantir-Piramesse – Stone and Metal Tools of Site Q I' in A. J. Veldmeijer and S. Ikram, eds, *Chasing Chariots: Proceedings of the First Chariot Conference* (Cairo 2012), Leiden (2013), 157–74.

41 See E. B. Pusch, *ÄgLev* 1, 103–5, fig. 12, pl. VII; *idem*, '"Pi-Ramesse-geliebt-von-Amun, Hauptquartier Deiner Streitwagentruppen": Ägypter un Hethiter in der Delta-Reziden der Ramessiden' in A. Eggebrecht, ed., *Pelizaeus-Museum Hildesheim: Die ägyptische Sammlung*, Mainz (1993), 136–7, fig. 135, 137–8; A. Herold, 'High Temperature Industries in the Late Bronze Age Capital Piramesse (Qantir)', 137–8, fig. 9–10; E. B. Pusch, 'Piramesse-Qantir: Residenz, Waffenschmiede und Drehscheibe internatinaler Beziehungen' in S. Petschel and M. von Falck, *Pharao siegt immer: Krieg und Frieden im alten Ägypten*, Bönen (2004), 242–4.

42 See Ch. Desroches-Noblecourt *et al.*, *Grand Temple d'Abou Simbel*, II, 27, pl. XXV–XXVI; A. Nibbi, 'Some Remarks on the Ancient Egyptian Shield', *ZÄS* 130 (2003), 179–80, fig. 5; J. Lorenz and I. Schrakamp, 'Hittite Military Warfare', 139, fig. 5.

43 T. D. Hulit, *Late Bronze Age Scale Armour*, 148.

44 E. B. Pusch and T. Rehen, *Hochtemperatur-Technologie in der Ramses-Stadt: Rubinglas für den Pharao*, Vol. 1–2, Hildesheim (2007), 822–4, Cat. GO 021-023.

45 E. B. Pusch, '"Pi-Ramesse-geliebt-von-Amun, Hauptquartier Deiner Streitwagentruppen": Ägypter un Hethiter in der Delta-Reziden der Ramessiden', 135.

46 For an overview of the characteristics of this type of headwear, see A. M. Snodgrass, *Arms and Armour of the Greeks*, 18–19; H. Hencken, *The Earliest European Helmets*, Cambridge, MA (1971), 18–20, fig. 3–4; P. Càssola Guida, *Le armi defensive*, 85–98; J. Borchhardt, 'Helme' in H.-G. Buchholz and J. Wiesner, *Kriegwesen, Teil 1: Schutzwaffen und Wehrbauten, Archaeologia Homerica* E 1, Göttingen (1977), 62, fig. 11; J. Aruz *et al.*, ed., *Beyond Babylon*, 440–2.

47 A. Herold, 'High Temperature Industries in the Late Bronze Age Capital Piramesse (Qantir): III. Workshops of The Temple of Amun Qantir/Piramesse, Site Q-I, Stratum B/2', 137–8; J. M. Kelder, 'Royal Gift Exchange between Mycenae and Egypt: Olives as "Greeting Gifts" in the Late Bronze Age Eastern Mediterranean',

AJA 113.3 (2009), 347, n. 61; *idem*, 'The Egyptian Interest in Mycenean Greece', *JEOL* 42 (2010), 127.

48 R. Parkinson and L. Schofield, 'Akhenaten's Army?', *EgArch*3 (1993), 34–5; *idem*, 'Of Helmets and Heretics: a Possible Egyptian Representation of Mycenaean Warriors on a Papyrus from El-Amarna', *ABSA* 89 (1994), 157–70, pl. 21–2; *idem*, 'Images of Mycenaeans: A Recently Acquired Painted Papyrus from El-Amarna' in W. V. Davies and L. Schofield, eds, *Egypt, the Aegean and the Levant: Interconnections in the Second Millennium BC*, London: British Museum (1995), 125–6; *idem*, 'A Painted Papyrus from Amarna' in J. Phillips, ed., *Ancient Egypt, the Aegean, and the Near East: Studies in Honour of Martha Rhoads Bell*, Austin (1998), 401–6.

49 J. D. S. Pendlebury, *The City of Akhenaten. Part III: The Central City and the Official Quarters. The Excavations at Tell el-Amarna during the Seasons 1926–1927 and 1931–1936*. Volume One: *Text*, London (1951), 140–1.

50 See N. De Garis Davies and A. H. Gardiner, *Tutankhamen's Painted Box*, Oxford (1962).

51 R. Parkinson and L. Schofield, *ABSA* 89, 160–1.

52 Beth Shan has been extensively excavated by the University Museum of the University of Pennsylvania from 1921 to 1933 and by the Hebrew University of Jerusalem in 1983 and from 1989 to 1996. For an overview of the role played by the town during the New Kingdom and the results of the excavations, see A. Mazar, 'Beth Shean in the Second Millennium BCE: From Canaanite Town to Egyptian Stronghold' in M. Bietak, ed., *The Synchronisation of Civilization in the Eastern Mediterranean in the Second Millennium BC II: Proceedings of the SCIEM 2000 – EuroConference, Haindorff, 2nd of May–7th of May 2001*, Vienna (2003), 322–33; *idem*, 'Tel Beth-Shean: History and Archaeology' in R. Kratz and H. Spieckermann, eds, *One God – One Cult – One Nation: Archaeological and Biblical Perspectives*, Berlin, New York (2010), 239–71; *idem*, 'The Egyptian Garrison at Beth-Shean' in S. Bar et al., eds, *Egypt, Canaan and Israel: History, Imperialism, Ideology and Literature. Proceedings of a Conference at the University of Haifa, 3–7 May 2009*, Leiden (2011), 151–85.

53 F. W. James and P. E. McGovern, *Late Bronze Egyptian Garrison*, 11, fig. 160.3; T. Hulit, *Late Bronze Age Scale Armour*, 141.

54 F. W. James and P. E. McGovern, *Late Bronze Egyptian Garrison*, 64, fig. 160.2, 4; T. Hulit, *Late Bronze Age Scale Armour*, 141.

55 F. W. James and P. E. McGovern, *The Late Bronze Egyptian Garrison at Beth Shean: A Study of Levels VII and VIII*, Philadelphia (1993), 15, fig. 160.1, T. Hulit, *Late Bronze Age Scale Armour*, 140–1.

56 See A. Mazar, *Excavations at Tel Beth-Shean 1989–1996. Vol. I: From the Late Bronze Age IIB to the Medieval Period*, Jerusalem (2006), 164, fig. 6.3, 11.

57 For an overview of this topic, see M. Burckhalter and A. Philippa – Touchais, 'Salamine', *BCH* 127, livraison 2 (2003), 729–33; Y. G. Lolos, 'Cypro-Mycenaean Relations ca. 1200 BC: Point Iria in the Gulf of Argos and Old Salamis in the Saronic Gulf' in N. C. Stampolidis, V. Karageorghis, eds, *ΠΛΟΕΣ. Sea Routes, Interconnections in the Mediterranean 16th–6th c. BC.: Proceedings of the International Symposium held at Rethymnon, Crete, September 29th–October 2nd 2002*, Athens (2003), 101–16; *ead.*, 'Salamis: Kanakia' in J. Whitley, ed., *Archaeology in Greece 2003–2004*, London (2004), 9–11; A. Philippa – Touchais, 'Salamine', *BCH* 128–29, livraison 2.2 (2004), 1299–1300; Y. G. Lolos, 'Salamis: Kanakia' in J. Whitley, ed., *Archaeology in Greece 2004–2005*, London (2005), 10; Y. G. Lolos *et al.*, 'Ajax's Capital. The Seat of the Maritime Kingdom of Salamis' in C. Pepe, ed., *Men, Lands and Seas: L'archeologia del mare. Atti del Convegno, Napoli 27–28 Giugno 2006*, Naples (2007), 114–27; N. Kourou, 'The Aegean and the Levant in the Early Iron Age: Recent Developments' in *Interconnections in the Eastern Mediterranean: Lebanon in the Bronze and Iron Ages. Proceedings of the International Symposium, Beirut 2008*, *BAAL* Hors-Série VI (2009), 362–3.

58 H. W. Catling, 'A Bronze Plate from a Scale-corslet found at Mycenae', *ArchAnz*, Heft 4 (1970), 441–9, fig. 1; *idem*, 'Panzer', 88, fig. 16a; P. Càssola Guida, *Le armi defensive*, 64.

59 J. Maran, 'The Spreading of Objects and Ideas in the Late Bronze Age Eastern Mediterranean: Two Case Examples from the Argolid of the 13th and 12th Centuries BC', *BASOR* 336 (2004), 18–19, fig. 14.

60 See P. Càssola Guida, *Le armi defensive*, 64; T. Cevoli, 'Rapporti militari tra mondo miceneo e Mediterraneo orientale alla luce di recenti scoperte archeologiche nell'isola di Salamina', in G. Borriello, ed., *Orientalia Parthenopea XV*, Napoli (2015), 68.

61 C. Spieser, *Les noms du Pharaon comme êtres autonomes au Nouvel Empire*, Fribourg, Göttingen (2000), 42–4, 48–51; *idem*, 'Cartouche', in E. Frood and W. Wendrich, eds, *UCLA Encyclopedia of Egyptology*, Los Angeles (2010), 5–6.

62 See J. von Beckerath, *Handbuch*, 154–7.

63 For an overview of the several variant orthographies of Ramesses II's *nomen*, see A. Spalinger, 'Historical Observations on the Military Reliefs of Abu Simbel and other Ramesside Temples in Nubia', *JEA* 66 (1980), 95–7; P. J. Brand, *The Monuments of Seti I: Epigraphic, Historical and Art Historical Analysis*, Leiden (2000), 35–6; C. Obsomer, *Ramsès II*, 65–7.

64 T. Cevoli, *Orientalia Parthenopea XV*, 74; R. D'Amato and A. Salimbeti, *Sea Peoples of the Bronze Age Mediterraean c. 1400 BC–1000 BC*, Oxford (2015), 23, 29–31.

65 See Chapter 5 'Terminology Relating to Protective Gear'.

66 H. Carter, *The Tomb of Tutankhamen Discovered by the Late Earl of Carnarvon and Howard Carter*, Vol. III, London (1933), 143.

67 The complete records of Howard Carter's discovery of the tomb of Tutankhamun are published on the website of the Griffith Institute – Oxford University with full and free access (www.griffith.ox.ac.uk/discoveringTut/).

68 We refer here to the essential article A. J. Veldmeijer et al., 'Tutankhamun's Cuirass Reconsidered', *JEOL* 48 (2021–2), 125–56.

69 A. J. Veldmeijer, *Amarna's Leatherwork. Part I: Preliminary Analysis and Catalogue*, Leiden (2011), 179; A. J. Veldmeijer et al., *JEOL* 48, 126, fig. 3.

70 G. Maspero, 'Le Papyrus Mallet', *RecTrav* 1 (1870), 47–59, pl. I–VI.

71 According to J. Janssen, the *deben* amounts to about 91 grams (J. J. Jenssen, *Commodity Prices from the Ramessid Period: An Economic Study of the Village of Necropolis Workmen of Thebes*, Leiden (1975), 101–2. In addition, it must be remembered Thomas Hulit provided an estimate of all the costs that a workshop incurred when manufacturing a body armour, relying on the data noted in the papyrus *Mallet* (T. D. Hulit, *Late Bronze Age Scale Armour*, 202–10).

72 *KRI* VI, 65, 7–8.

5 Egyptian Terminology Relating to Protective Gear

1 *Urk* IV, 711, 8–712, 1; PM II, 89 [240]–[244]. This Asiatic campaign was fought during regnal year 35 of Thutmose III. See D. B. Redford, *The Wars in Syria and Palestine of Thutmose III*, Leiden, Boston (2003), 83–5.

2 For example, ⌒𓊽𓏺𓏤𓊖 *dbn* noun, 'ring, circle' (*Wb* V, 436, 6-9); ⌒𓊽𓊖 *dbn* verb, 'to be wrapped' (*Wb* V, 436, 12); ⌒𓊽𓊽𓂻 *dbn* verb, 'to go around in circles' (*Wb* V, 439, 1). The latter is distinguished by the reduplication of the triliteral root *dbn*, following the ABCBC pattern (A. H. Gardiner, *Egyptian Grammar*, 3rd edn, Oxford (2001), 210–11; P. Vernus, 'Égyptien', *École pratique des hautes études. 4e section, sciences historiques et philologiques. Livret 4. Rapports sur les conférences des années 1985–1986 & 1987*, Paris (1994), 194; A. Loprieno, *Ancient Egyptian: A Linguistic Introduction*, Cambridge (1995), 54; J. Allen, *Ancient Egyptian Phonology*, Cambridge (2020), 123–5.

3 Sydney Aufrère relates that determinative with 'la représentation d'un outil ou d'une partie d'outil traditionnellement consitué de métal, une lame de couteau, un fer de hache'. See S. Aufrère, *L'Univers Minéral dans la pensée égyptienne*, Vol. 1, Cairo (1991), 106.

4 For example, 𓂝𓈎𓏏𓏤 *ikḥw*, 'axe' (*Wb* I, 138, 18); 𓃀𓎼𓋴𓅱 *bꜣgsw*, 'dagger' (*Wb* I, 432, 5); 𓏎𓈖𓃀 *minb*, 'axe' (*Wb* II, 44, 8); 𓏶𓏏 *mtf.t*, 'dagger' (*Wb* II, 170, 7); 𓏠𓈖𓏏 *mtni.t*, 'axe' (*Wb* II, 171, 6); 𓎛𓂝𓅱 *ḥꜥw*, 'weapons' (*Wb* III, 242, 11–14); 𓋴𓆑𓏏 *sf.t*, 'sword' (*Wb* IV, 442, 8); 𓎡𓂋𓂧𓈖 *krḏn*, 'axe' (*Wb* V, 66, 7).

Moreover, we can add to the previous list 𓇋𓂝𓅓𓏛𓏤, an unusual writing of the word *ikm* (*Wb* I, 139, 13–15) featuring the determinative 𓂝. That peculiar writing of the word *ikm* appears uniquely in one place in the *Annals* of Thutmose III (*Urk*, IV, 719, 1).

5 See M. Liverani, *Guerra e diplomazia*, 208–9.

6 See J. R. Harris, *Lexicographical Studies in Ancient Egyptian Minerals*, Berlin (1961), 63–4; E. Delange, *Monuments égyptiens du Nouvel Empire: la Chambre des Ancêtres, les Annales de Thoutmosis III et le décor de(s) palais de Séthi Ier*, Paris (2015), 158.

7 A. Loprieno, *Ancient Egyptian*, 38; J. Allen, *The Ancient Egyptian Language: An Historical Study*, Cambridge (2013), 23, 54; J. Allen, *Phonology*, 40.

8 The so-called 'Cycle of Inaros/Petubastis' is known through extant texts spanning the period from 500 BCE to 200 CE. As stated by Kim Ryholt the cycle 'constitutes the largest, connected group of narrative literature from ancient Egypt' (K. Ryholt, 'The Assyrian Invasion of Egypt in Egyptian Literary Tradition' in J. G. Dercksen, ed., *Assyria and Beyond: Studies Presented to Mogens Trolle Larsen*, Leiden (2004), 491). For a general overview of the stories of the Cycle of Inaros already published, see M. Chauveau, 'Les richesses méconnues de la littérature démotique', *BSFE* 156 (2003), 31–5; M. Chauveau, 'Les romans du cycle d'Inaros et de Pédoubastis', *Egypte* 29 (2003), 19–28; D. Agut-Labordère and M. Chauveau, *Héros, magiciens et sages oubliés de l'Égypte ancienne*, Paris (2011), 67–143.

9 See W. Spiegelberg, 'Demotische Miscellen', *RecTrav* 30 (1908), 154; ead., *Der Sagenkreis des Königs Petubastis* (*DemStud* 3), Leipzig (1910), 16,18, 4,15, 66*, 458; E. Bresciani, *Letteratura*, 913, 921; F. Hoffmann and J. F. Quack, *Anthologie der demotischen Literatur*, Berlin (2007), 95, 105; D. Agut-Labordère and M. Chauveau, *Héros*, 79, 93.

10 See A. Volten, *Ägypter und Amazonen: Eine demotische Erzählung des Inaros-Petubastis-Kreises aus zwei Papyri der Österreichen Nationalbibliothek (Pap. Dem. Vindob. 6165 und 6165 A)*, (*MPER* Neue Serie 6), Vienna (1962), 28–9; F. Hoffmann, *Ägypter und Amazonen: Neubearbeitung Zweier demotischer Papyri P. Vindob D 6165 und P. Vindob D 6165 A*, (*MPER* Neue Serie 24), Vienna (1995), 48–9; D. Agut-Labordère and M. Chauveau, *Héros*, 135.

11 See above section 2.2 'Helmets and armour as prizes of war'.

12 See D. B. Redford, *Wars*, 35, n. 207.

13 *KRI* II, 290.3; PM IV, 21 [196]; J. Yoyotte, 'Les Stèles de Ramsès II a Tanis: Première Partie', *Kêmi* 10 (1949), 67; J. P. Emanuel, '"Šrdn from the Sea": The Arrival, Integration, and Acculturation of a "Sea People"', *Journal of Ancient Egyptian Interconnections* 5.1 (2013), 15.

14 *Urk* IV, 664, 3–5; see also *LD* III, pl. 32; H. K. Brugsch, *Thesaurus inscriptionum Aegyptiacarum: altaegyptische Inschriften (Band 5): Historisch-biographische*

Inschriften altaegyptischer Denkmaeler, Leipzig (1891), 1164, §15; J. H. Breasted, *Ancient Records*, 187, §435; J. A. Wilson, 'Egyptian Historical Texts' in J. B. Pritchard, ed., *ANET*, Princeton, NJ (1954), 237; D. B. Redford, *Wars*, 35, n. 207–8.

15 *Urk* IV, 664, 7. See also *LD* III, pl. 31 [a]; H. K. Brugsch, *Thesaurus*, 1178, §40–41; J. H. Breasted, *Ancient Records*, 208, §§500–501, 210, §509; E. Delange, *Monuments*, 144–5.

16 *Urk* IV, 726, 17, 732, 1; *LD* III, pl. 30 [a]; H. K. Brugsch, *Thesaurus*, 1182, §5, 1184, §15; J. H. Breasted, *Ancient Records*, 213, §525, 216, §534.

17 *Urk* IV, 1235, 9; see also G. A. Reisner, 'Inscribed monuments from Gebel Barkal', *ZÄS* 66 (1931), 80, (2); G. A. Reisner and M. B. Reisner, 'Inscribed monuments from Gebel Barkal, Part 2: The Granite Stela of Thutmose III', *ZÄS* 69 (1933), 32; J. A. Wilson, *ANET*, 238; J. K. Hoffmeier, 'The Gebel Barkal Stela of Thutmose III (2.2 B)' in W. W. Hallo and K. L. Younger, eds, *The Context of Scripture*, Vol. 2: *Monumental Inscriptions from the Biblical World*, Leiden, Boston, Köln (2000), 16.

18 *Urk*, IV, 1311, 16; see also G. Legrain, *ASAE* 4, 126–31; E. Edel, *ZDPV* 69, 97–126; J. K. Hoffmeier, 'The Memphis and Karnak Stelae of Amenhotep II' in W. W. Hallo and K. L. Younger, eds, *The Context of Scripture*, Vol. 2: *Monumental Inscriptions from the Biblical World*, Leiden, Boston, Köln (2000), 20.

19 It has been suggested that the word 𓅓𓋴𓋴𓏏 *mss(.t)* could designate a bag-tunic resembling a modern *galabeya*. See J. J. Janssen, *Commodity Prices*, 260–1; R. Hall, 'The Pharaonic mss tunic ([𓅓𓋴𓋴𓏏) as a smock', *GöttMisz* 43 (1981), 29–37; G. Vogelsang-Eastwood, *Pharaonic Egyptian Clothing*, 8–9, 130, n. 1.

20 W. Ventzke, 'Zur Rekonstruktion eines bronzenen Schuppenpanzer', 97.

21 See A. Roccati, 'Il bilinguismo interno dell'Egitto', *VicOr* 3 (1980), 77–80; ead., 'La lingua diffusa (politica e lingua nell'Egitto ramesside)', *ParPass* 268 (1993), 26–37.

22 See W. Helck, *Die Beziehungen Ägyptens*, 505–75. This phenomenon of lexical borrowing diminished considerably after the Ramesside Period, following the decline of Egyptian rule in Western Asia. In this regard, see J. E. Hoch, *Semitic Words*, 3–5; O. El Aguizy and F. Haykal, 'Changes in Ancient Egyptian Language', *EgMa* 27–8 (1996), 31–2.

23 C. G. Brandenstein, 'Zum Churrischen Lexicon', *ZeitAss* 46 (1940), 104 *et seqq.*; E. A. Speiser, 'On some Articles of Armor and their Names', *JAOS* 70.1 (1950), 47; E. Laroche, *Glossaire de la langue hurrite*, Paris (1980), 215–16; T. Richter, *Bibliographisches Glossar des Hurritischen*, Wiesbaden (2012), 357; T. Dezsö, *RLA* 10 (2003–5), 321.

24 R. H. Beal, 'I reparti e le armi dell'esercito ittita' in M. C. Guidotti and F. Pecchioli Daddi, eds, *La battaglia di Qadesh: Ramesse II contro gli Ittiti per la conquista della Siria*, Livorno (2002), 95; T. Dezsö, *RLA* 10 (2003–5), 321; G. Güterbock *et al.*, eds,

The Hittite Dictionary of the Oriental Institute of the University of Chicago, Vol. Š, fasc. 2, Chicago (2005), 259.
25 *ADOI* 15, S, 313–14; J. E. Hoch, *Semitic Words*, 366 [546].
26 E. A. Speiser, *JAOS* 70.1, 47–8; J. E. Hoch, *Semitic Words*, 637. In contrast to the limited amount of Egyptian data available, the Akkadian texts found in Nuzi (modern Yorghan Tepe, Iraq), dating roughly from the middle of the fifteenth to the mid-fourteenth century BCE, list no fewer than sixteen different types of helmets, corslets and their components. See E. R. Lacherman, 'Epigraphic Evidences of the Material Culture of the Nuzians' in R. F. S. Starr, *Nuzi: Report on the Excavations at Yorgan Tepa near Kirkuk, Iraq, conducted by Harvard University in Conjunction with the American School of Oriental Research and the University Museum of Philadelphia*, Vol. I, Cambridge, MA (1939), 540–1; T. Kendall, 'gurpisu ša awēli: The Helmets of the Warriors of Nuzi', 200–4.
27 T. Dezsö, *RLA* 10 (2003–5), 321; J. E. Hoch, *Semitic Words*, 367.
28 E. A. Speiser, *JAOS* 70.1, 47; T. Dezsö, *RLA* 10 (2003–5), 321. See also discussion in J. R. Zorn, *BASOR* 360, 3–6.
29 A. A.-H. Youssef, 'A Nineteenth Dynasty New Word for Blade and the Semitic Origin of Some Egyptian Weapon-Names and Other Related Words', *MDAIK* 39 (1983), 259; J. E. Hoch, *Semitic Words*, 367.
30 According to Timoty Kendall, the language used in the Tell el-Amarna tablets was '... an artificial Babylonian (Akkadian) dialect, liberally infused with West Semitic and Egyptian words ...' (T. Kedall, 'Foreign Relations' in R. E. Freed *et al.*, eds, *Pharaohs of the Sun: Akhenaten, Nefertiti, Tutankhamen*, London (1999), 157–8). On this aspect, see also M. Müller, *Akkadisch in Keilschrifttexten aus Ägypten* (*AOAT* 373), Münster (2010); W. H. van Soldt, 'Akkadian as a Diplomatic Language' in S. Weninger, ed., *The Semitic Languages: An International Handbook*, The Hague (2011), 405–15.
31 P. Collombert and L. Coulon, *BIFAO* 100, 193–242.
32 See A. H. Gardiner, *Late Egyptian Stories*, Bruxelles (1932), XII, 76–81; ead. 'The Astarte Papyrus' in S. R. K. Glanville, ed., *Studies presented to F. Ll. Griffith*, London (1932), 74–85, pl. 8–9; A. H. Sayce, *JEA* 19.1/2, 56–9.
33 P. Collombert and L. Coulon, *BIFAO* 100, 193–9.
34 P. Vernus, 'Réception linguistique et idéologique d'une nouvelle technologie: le cheval dans la civilisation pharaonique' in M. Wissa, ed., *The knowledge economy and technological capabilities of Egypt, the Near East and the Mediterranean, 2nd millennium BC–1st millennium AD: proceedings of a confererence held at the Maison de la Chimie, Paris, France 9–10 December 2005*, Sabadell (2010), 1–46.
35 There is no agreement among scholars regarding the etymology of the Egyptian word *ssm.t*. Two main theories have emerged concerning its origin. The first theory suggests that the origin of the word must be sought in the Proto-Indo-

European. The second theory suggests that the origin of the word can be traced back to the Sumerian. See S. Turner, *The Horse*, 27-9.

36 H. Altenmüller and A. M. Moussa, 'Die Inschrift Amenemhets II. aus dem Ptah - Tempel von Memphis: Ein Vorbericht', *SAK* 18 (1991), 1-46.

37 P. Collombert and L. Coulon, *BIFAO* 100, 215.

38 T. Schneider, 'Fremdwörter in der ägyptischen Militärsprache der Neuen Reiches und ein Bravourstück des Elitesoldaten (Papyrus Anastasi I 23, 2-7)', *JSSEA* 35 (2008), 192 [54].

39 *KRI* II, 28.7; C. Kuentz, *Bataille*, 29, 237, §77.

40 *KRI* II, 28.9; C. Kuentz, *Bataille*, 76, 237, §77.

41 *KRI* II, 119.12; C. Kuentz, *Bataille*, 100, 354, §86.

42 *KRI* II, 28.10; C. Kuentz, *Bataille*, 121, 237, §77.

43 *KRI* II, 119.13; C. Kuentz, *Bataille*, 142, 354, §86.

44 *KRI* II, 175. 3-12.

45 *KRI* II, 119.14; C. Kuentz, *Bataille*, 161, 354, §86.

46 *KRI* II, 174.4-12.

47 *KRI* II, 119,15; C. Kuentz, *Bataille*, 188, 354, §86.

48 *KRI* II, 28.11-13; A. H. Gardiner, *Hieratic Papyri*, 23-4, pl. 9-9a, 10-10a.

49 *KRI* II, 28.12; E. de Rougé, *RecTrav* I, 1-9 + facsimile; C. Kuentz, *Bataille*, 237, §77.

50 J. Allen, *Egyptian Language*, 31-3.

51 A. Loprieno, *Ancient Egyptian*, 32-3; J. Allen, *Egyptian Language*, 31-3; J. Allen, *Phonology*, 42-3.

52 F. Junge, *Late Egyptian Grammar: An Introduction*, Oxford (2005), 41; J. Allen, *Phonology*, 36-7.

53 For an in-depth analysis of 'group writing', see M. Burchardt, *Die altkanaanäischen Fremdworte und Eigennamen in Ägyptischen*, Leipzig (1910), I2-I6, §§ 3-10; W. Albright, *The Vocalisation of the Egyptian Syllabic Orthography*, AOS 5, New Haven, CT (1934), 1-30; J. Zeidler, 'A New Approach to the Late Egyptian "Syllabic Orthography"' in *Sesto Congresso Internazionale di Egittologia, Atti*, Vol. II, Turin (1993), 579-90; F. Neveu, *La langue des Ramsès: Grammaire du néo-égyptien*, Paris (1998), 307-10; J. E. Hoch, *Semitic Words*, 485-504; F. Junge, *Late Egyptian*, 41-5; J. Allen, *Phonology*, 36-8.

54 W. Albright, *Vocalisation*, 36, IV, 15.

55 T. Kendall, *Warfare and Military Matters*, 263.

56 T. Dezsö, 'Scale Armour of the 2nd Millennium BC' in T. A. Bács, ed., *A Tribute to Excellence: Studies Offered in Honor of Ernő Gaál, Ulrich Luft, László Török* (Studia Egyptiaca XVII), Budapest (2002), 196, note 4.

57 S. Hassan, *The Sphinx: Its History in the Light of Recent Excavations*, Cairo (1949), 150-1, fig. 34; idem, *Excavations at Gîza 8: 1936-1937. The Great Sphinx and Its*

Secrets: Historical Studies in the Light of Recent Excavations, Cairo (1953), 259–60, fig. 195; C. M. Zivie, Giza au deuxième millénaire, Cairo (1976), 235–7.
58 PM III/1, 46.
59 For an overview of the goddess Isis and her worship, see R. Krauß, 'Isis', LÄ, Band 3 (1980), 186–203; J.-P. Corteggiani, L'Égypte ancienne et ses dieux: Dictionnaire illustré, Paris (2007), 244–9.
60 See F. B. Anthony, Foreigners in Ancient Egypt, 25–6, fig. 9.
61 S. Hassan, The Great Sphinx, 260; R. Stadelmann, Syrisch – Palästinensische Gottheiten in Ägypten, Leiden (1967), 124; C. M. Zivie, Deuxième millénaire, 236; idem, Giza au premier millénaire: Autour du temple d'Isis Dame des Pyramides, Boston (1991), 24–5.
62 See V. Vikentiev, 'Les rites de la réinvestiture royale en tant que champ de recherches sur la période archaïque égypto-libyenne', BIE 37 (1957), 293–4, pl. III.
63 A. Radwan, 'The First Appearance of Isis in a Foreign Dress', Memnonia IX (1998), 176.
64 For an overview of the role played by Shaushka in Hurrian mythology, see G. Gestoso Singer, 'Shaushka, the Traveling Goddess', TrabEg 6 (2016), 44–8; M. R. Bachvarova, 'Adapting Mesopotamian Myth in Hurro-Hittite Rituals at Hattuša; Ištar, the Underworld and the Legendary Kings ' in B. J. Collins and P. Michalowski, eds, Beyond Hatti: A Tribute to Gary Beckman, Atlanta (2013), 24–9.
65 See V. Haas, Geschichte der hethitischen Religion, Leiden, New York, Köln (1994), 349–50; G. Beckman, 'Ištar of Nineveh Reconsidered', JCunStud 50 (1998), 3–4.
66 V. Haas, Religion, 350–1; M. C. Trémouille, 'Šauška, Šawuška', RLA 12, Berlin, New York (2009), 101.
67 For an overview of the spreading of the Hurrian language in the Hittite world during the Late Bronze Age, see S. de Martino, 'The Hurrian Language in Anatolia during the Late Bronze Age' in A. Mouton, ed., Hittitology Today: Studies on Hittite and Neo-Hittite Anatolia in Honor of Emmanuel Laroche's 100th Birthday: 5e Rencontres d'archéologie de l'IFEA, Istanbul 21–22 novembre 2014, Istambul (2017), 151–62.
68 Šamuha was one of the main cult centres of Shawshka. At the end of the fifteenth century BCE, a statue of the goddess was moved from the kingdom of Kizzuwatna to the sanctuary of the town. See H. Haas, Religion, 578–80; G. Beckman, JCunStud 50, 4. For the alleged location of Šamuha, see S. de Martino, 'The Hittite City of Šamuḫa: Its Location and Its Religious and Political Role in the Middle Kingdom' in K. Strobel, ed., New Perspectives on the Historical Geography and Topography of Anatolia in the II and I Millennium BC, Florence (2008), 131–43.
69 I. Wegner, Gestalt und Kult der Ištar-Šawška in Kleinasien, Neukirchen-Vluyn (1981), R. H. Beal, The Organisation of the Hittite Military, Heidelberg (1992),

150-1, notes 549, 550; *idem.*, 'I reparti e le armi dell'esercito ittita', 95-6; I. Wegner, *Corpus der hurritischen Sprachdenkmäler (ChS), I. Abteilung: Die Texte aus Boğazköy, Band 3. Hurritische Opferlisten aus hithitischen Festbeschreibungen, Teil. 1 Texte für Ištar – Ša(w)uška*, Rome (1995), 39, 42, 70.

70 See T. Schneider, *JSSEA* 35, 191 [40]; J. Winand, 'Identifying Semitic Loanwords in the Late Egyptian', 486-9.

71 See J. E. Hoch, *Semitic Words*, 202-3 [274].

72 F. Junge, *Late Egyptian Grammar*, 42-3.

73 See A. Wiedemann, *Hieratische Texte aus den Museen zu Berlin und Paris*, Leipzig (1879), 19-23, pl. X-XIV; A. Erman and F. Krebs, *Aus den Papyrus der Königlichen Museen*, Berlin (1899), 93-7; A. H. Gardiner, *Egyptian Hieratic Texts*, 36*-38*, 42, 1.

74 In the context of the late Egyptian miscellanies, the 'list' genre is quite widespread. The use of this kind of literary work enabled the author, on the one hand, to give a certain symmetry to his text, and on the other, to show the richness of his vocabulary. Moreover, the strong presence of loan words can be seen as a further attempt to highlight the author's erudition. See C. Ragazzoli, 'Les manuscripts de miscellanies en Égypte ancienne, ou la lecture comme pratique créative' in S. Morlet, ed., *Lire en extraits. Une contribution à l'histoire de la lecture et de la literature, de l'Antiquité au Moyen Âge*, Paris (2015), 21-3.

75 In Pap. *Koller*, many of the technical terms which concern the parts of the chariot and its panoply also occur in the ostraca National Museum of Scotland A. 1956.319 (formerly Edinburgh Ostracon (O.). 916) and Turin S 9588 (formerly CG 57365). The two ostraca provide two different fragments of the so-called 'Hymn to the King in his Chariot', a lengthy poetic composition celebrating the victorious pharaoh riding his chariot. See C. Manassa, 'The Chariot that Plunders Lands: "The Hymn to the King in his Chariot"' in A. J. Veldmeijer and S. Ikram, eds, *Chasing Chariots*, Leiden (2013), 21-3.

76 The Demotic papyrus P. Carlsberg 80 is one of the five manuscripts from Tebtunys containing the *Inaros Epic* (P. Carlsberg 68 + 123, P. Carlsberg 80, P. Carlsberg 164, P. Carlsberg 458 and P. Carlsberg 591). See E. Bresciani, 'La corazza di Inaro era fatta con la pelle del grifone del Mar Rosso', *EVO* 13 (1990), 103-7; *ead.*, *Letteratura*, 945-7; K. Ryholt, 'The Assyrian Invasion of Egypt in Egyptian Literary Tradition', 492-5.

77 See W. Spiegelberg, *Sagenkreis*, 11-42; E. Bresciani, *Letteratura*, 909-21; F. Hoffmann and J. F., Quack, *Anthologie*, 96-106; D. Agut-Labordère and M. Chauveau, *Héros*, 71-94.

78 See J. Krall, 'Der demotische Roman aus der Zeit des Königs Petubastis', *WZKM* 17 (1903), 15-33; W. Spiegelberg, *Sagenkreis*, 43-75; E. Bresciani, *Der Kampf um*

den Panzer des Inaros (Papyrus Krall), Vienna (1964), 50–109; F. Hoffmann, *Der Kampf um der Panzer des Inaros: Studien zum P. Krall und seiner Stellung innerhalb des Inaros-Petubastis Zylus*, Vienna (1996), 229–389; E. Bresciani, *Letteratura*, 922–40; F. Hoffmann and J. F. Quack, *Anthologie*, 70–86; D. Agut-Labordère and M. Chauveau, *Héros*, 95–132.

79 See A. Volten, *Ägypter und Amazonen*, 28–65; F. Hoffmann, *Ägypter und Amazonen*, 49–117; M. Chauveau, *Egypte* 29, 23–6; F. Hoffmann and J. F. Quack, *Anthologie*, 110–16; D. Agut-Labordère and M. Chauveau, *Héros*, 133–43.

80 See W. E. Crum, *A Coptic Dictionary*, Oxford (1939), 630, 668; J. Černý, *Coptic Etymological Dictionary*, Oxford (1976), 70; W. Vycichl, *Dictionnaire étymologique de la langue copte*, Leuven (1983), 297; J. E. Hoch, *Semitic Words*, 202.

81 For an exhaustive survey of the Semitic loan words used in Egyptian and survived until Coptic, see C. Peust, *Egyptian Phonology. An Introduction to the Phonology of a Dead Language*, Göttingen (1999), 307–9.

82 Cfr. John of Constantinople Discourse, *On Penitence and Abstinence*, part 1 (p. CMCL.AW218), Chapter 41: (1) ⲉⲧⲃⲉⲛⲁⲓ ⲧⲏⲣⲟⲩ ⲉⲣⲉⲡⲙⲁⲕⲁⲣⲓⲟⲥ ⲡⲁⲩⲗⲟⲥ ⲱⲣⲝ ⲙⲙⲟⲛ· (2) ⲁϥⲥϩⲁⲓ ϫⲉⲭⲓ ⲛⲏⲧⲛ ⲛⲧⲡⲁⲛϩⲟⲡⲗⲓⲁ ⲙⲡⲛⲟⲩⲧⲉ ·ⲙⲛⲧⲡⲉⲣⲓⲕⲉⲫⲁⲗⲁⲓⲁ ⲙⲡⲉⲡⲛⲉⲩⲙⲁ·

'(1) Because of all these things, the blessed man Paulus armed us with weapons, (2) for he wrote saying: "Take unto you the whole armour of God, and the helmet of the Spirit"'.

83 See J. E. Hoch, *Semitic Words*, 203 [275].

84 For the nature of the Domain of Amun and the aspects related to its organization from the end of the New Kingdom to the early Third Intermediate Period, see A. Gasse, *Données nouvelles administratives et sacerdotales sur l'organisation du domaine d'Amon. XXe-XXIe dynasties*, Cairo (1988), 171–233; J.-C. Antoine, 'The Wilbour Papyrus Revisited: The Land and Its Localization. An Analysis of the Places of Measurement', *SAK* 40 (2011), 9–27.

85 A. Gasse, *Données nouvelles*, 81, 85 (30), pl. 40, 140.

86 See G. Sauneron, *Le papyrus magique illustré de Brooklyn [Brooklyn Museum 47.218.156]*, New York (1970), 24, 27, pl. V.

87 *Ivi*, 12–13, fig. 1.

Bibliography

Abbas, M. R., 'The Maryannu in the Western Desert during the Ramesside Period', *Abgadiyat* 8 (2013), 127–32.

Abbas, M. R., 'The Bodyguard of Ramesses II and the Battle of Kadesh', *ENiM* 9 (2016), 113–23.

Abbas, M. R., 'A Survey of the Military Role of the Sherden Warriors in the Egyptian Army during the Ramesside Period', *ENiM* 10 (2017), 7–23.

Abbas, M. R., 'The Town of Yenoam in the Ramesside War Scenes and Text of Karnak', *CahKarn* 16 (2017), 329–41.

Abdul-Qader, M., 'Preliminary Report of the Excavations carried out in the temple of Luxor, Seasons 1958–1959 and 1959–1960', *ASAE* 60 (1968), 227–79.

Abrahami, P., 'L'armée d'Akkad' in P. Abrahami and L. Battini, eds, *Le armées du Proche-Orient ancient: IIIe–Ier av. J. C. Actes du Colloque International Organisé à Lyon les 1er et 2 décembre 2006, Maison de l'Orient et de la Méditerranée* (BAR-IS 1855), Oxford: BAR Publishing (2008), 1–20.

Agut-Labordère, D. and Chauveau, M., *Héros, magiciens et sages oubliés de l'Égypte ancienne*, Paris: Les Belles Lettres (2011).

Albright, W. F., 'Mitannian maryannu, "chariot-warrior", and the Canaanite and Egyptian Equivalent', *AOF* 6 (1930–1), 217–21.

Albright, W. F., *The Vocalisation of the Egyptian Syllabic Orthography* (AOS 5), New Haven, CN: American Oriental Society (1934) (Kraus Reprint Co, Millwood, 1974).

Aldred, C., 'The Foreign Gifts Offered to Pharaoh', *JEA* 56 (1970), 105–16.

Allen, J., *The Ancient Egyptian Language: An Historical Study*, Cambridge: Cambridge University Press (2013).

Allen, J., *Ancient Egyptian Phonology*, Cambridge: Cambridge University Press (2020).

Allon, N., 'Seth is Baal–Evidence from the Egyptian Script', *ÄgLev* 17 (2007), 15–22.

Alster, B., 'Images and Text on the "Stele of the Vultures"', *AOF* Bd. 50 (2003/4), 1–10.

Altenmüller, H. and Moussa, A. M., 'Die Inschrift Amenemhets II. aus dem Ptah – Tempel von Memphis. Ein Vorbericht', *SAK* 18 (1991), 1–48.

Amiet, P., *L'art de l'Agadé au musée du Louvre*, Paris: Réunion des Musées nationaux (1976).

Andreu-Lanoë, G., Labbé-Toutée, S. and Rigault, P., eds, *L'art du contour: Le dessin dans l'Égypte ancienne*, Paris: Somogny Editions (2013).

Andrikou, E., 'New Evidence on Mycenaean Bronze Corselets from Thebes in Boeotia and the Bronze Age Sequence of Corselets in Greece and Europe' in I. Galanaki, H. Tomas, Y. Galanakis and R. Laffineur, eds, *Between the Aegean and Baltic Seas: Prehistory across Borders : Proceedings of the International Conference Bronze and Early Iron Age Interconnections and Contemporary Developments between the Aegean and the Regions of the Balkan Peninsula, Central and Northern Europe, University of Zagreb, 11–14 April 2005*, Universitè de Liège, Histoire de l'art et archéologie de la Grèce antique; University of Texas at Austin, Program in Aegean Scripts and Prehistory, Liège, Belgique, Austin, TX (2007), 401–9.

Angenot, V., 'Les peintures de la chapelle de Sennefer (TT 96A)', *Egypte* 45 (2007), 21–32.

Anthony, F. B., *Foreigners in Ancient Egypt: Theban Tomb Paintings from the Early Eighteenth Dynasty*, London, New York: Bloomsbury (2017).

Antoine, J.-C., 'The Wilbour Papyrus Revisited: The Land and Its Localisation: An Analysis of the Places of Measurement', *SAK* 40 (2011), 9–27.

Archer, R., 'Chariotry and Cavalry: Developments in the Early First Millennium' in G. G. Fagan and M. Trundle, eds, *New Perspectives on Ancient Warfare*, Leiden, Boston: Brill (2010), 57–79.

Arnold, D., *Der Temple des Königs Mentuhotep von Deir el-Bahari, Band III: Die königlichen Beigaben*, Mainz: Philipp von Zabern (1981).

Arnold, D., *The Pyramid of Senwosret I (MMAEE Vol. 21)*, New York: Metropolitan Museum of Arts (1988).

Aruz, J. and Wallenfels, R., eds, *Art of the First Cities: The Third Millennium BC from the Mediterranean to the Indus*, New York: Metropolitan Museum of Arts (2003).

Aruz, J., Benzel, K. and Evans, J. M., ed., *Beyond Babylon: Art, Trade, and Diplomacy in the Second Millenium BC*, New York: Metropolitan Museum of Arts (2008).

Astour, M. C., 'The Partition of the Confederacy of Mukiš – Nuḫašše–Nii by Šuppiluliuma: A Study in Political Geography of the Amarna Age', *Orientalia* 39 (1969), 381–414.

Astour, M. C., 'Tunip-Hamat and Its Region', *Orientalia* 46 (1977), 51–64.

Aufrère, S., *L'Univers Minéral dans la pensée égyptienne*, Vol. 1 (*BiEtud* 105/1), Cairo: IFAO (1991).

Aufrère, S., 'Les vétérans de Montouhotep Nehépetrê: Une garnison funéraire à Deir al-Bahari?', *Egypte* 19 (2000), 8–16.

Ayali-Darshan, N., *Storm-God and the Sea: The Origin, Version, and Diffusion of a Myth throughout the Ancient Near East*, Tübingen: Mohr Siebeck (2020).

Bachvarova, M. R., 'Adapting Mesopotamian Myth in Hurro-Hittite Rituals at Hattuša; Ištar, the Underworld and the Legendary Kings' in B. J. Collins and P. Michalowski, eds, *Beyond Hatti: A Tribute to Gary Beckman*, Atlanta: Lockwood Press (2013), 23–44.

Badawy, A. M., 'Die neue historische Stele Amenophis' II', *ASAE* XLII (1943), 1–23.

Barbotin, C., *Âhmosis et le début de la XVIIIe dynastie*, Paris: Pygmalion (2008).

Bermond Montanari, G., 'La tomba di Carpena (Forlì)' in G. Bermond Montanari, M. Massi Pasi and L. Prati, eds, *Quando Forlì non c'era: Catalogo della mostra*, Forlì: A.B.A.C.O. (1997), 273–7.

Barrelet, M.-T., 'Peut-On Remettre en Question la "Restitution Matériel de la Stèle des Vautours"?', *JNES* 29.4 (1970), 233–58.

Basmachi, F., 'An Akkadian Stele', *Sumer* 10 (1954), 116–19.

Beal, R. H., *The Organisation of the Hittite Military*, Heidelberg: Carl Winter-Universitätsverlag (1992).

Beal, R. H., 'I reparti e le armi dell'esercito ittita' in M. C. Guidotti and F. Pecchioli Daddi, eds, *La battaglia di Qadesh: Ramesse II contro gli Ittiti per la conquista della Siria*, Livorno: Sillabe (2002), 93–108.

Beckman, G., 'Ištar of Nineveh Reconsidered', *JCunStud* 50 (1998), 1–10.

Bell, L., 'The Epigraphic Survey', *OIR 1986–87* (1987), 4–17.

Bell, L., 'The Epigraphic Survey', *OIR 1988–89* (1990), 4–8.

Ben-Tor, D., ed., *Pharaoh in Canaan: The Untold Story*, Jerusalem: Israel Museum (2016).

Bestock, L., *Violence and Power in Ancient Egypt: Image and Ideology before the New Kingdom*, London, New York: Routledge (2018).

Bietak, M., 'Harbours and Coastal Military Bases in Egypt in the Second Millennium BC: Avaris, Peru-nefer, Pi-Ramesse' in H. Willems and J.-M. Dahms, eds, *The Nile: Natural and Cultural Landscape in Egypt*, Bielefeld: Transcript Verlag (2017), 53–70.

Bisson de la Roque, F., *Rapport sur les Fouilles de Médamoud. Année 1929 (FIFAO 7)*, Cairo (1930).

Blackman A. M. and Peet, T. E., 'Papyrus Lansing: A Translation with Notes', *JEA* 11 (1925), 284–98.

Bonnet, H., *Die Waffen der Völker des alten Orients*, Leipzig: Hinrichs (1926).

Boraik, M., 'Sphinxes Avenues Excavation: First Report', *CahKarn* 13 (2010), 45–78.

Boraik, M., 'The Sphinxes Avenue Excavations to the East Bank of Luxor' in M. G. Folli, ed., *Sustainable Conservation and Urban Regeneration: The Luxor Example*, Cham: Springer (2018), 7–31.

Borchardt, L., *Statuen und Statuetten von Königen und Privatleuten im Museum von Kairo, Nr. 1-1294. Teil I, Text und Tafeln zu Nr. 1-380* (CGC n. 1–1294), Berlin: Reichsdruckerei (1911).

Borchardt, J., 'Helme' in H.-G. Buchholz and J. Wiesner, *Kriegwesen, Teil 1: Schutzwaffen und Wehrbauten, Archaeologia Homerica* E 1, Göttingen: Vandenhoeck & Rupprecht (1977), 57–74.

Brand, P. J., *The Monuments of Seti I: Epigraphic, Historical and Art Historical Analysis* (ProblÄg 16), Leiden: Brill (2000).

Brandenstein, C. G., 'Zum Churrischen Lexicon', *ZeitAss* 46 (1940), 83–115.

Breasted, J. H., *The Battle of Kadesh: A Study in the Earliest Known Military Strategy*, Chicago: The University of Chicago Press (1903).

Breasted, J. H., *Ancient Records of Egypt from the Earliest Times to the Persian Conquest, Collected Edited and Translated with Commentary*, Vol. II, *The Eighteenth Dynasty*, Chicago: The University of Chicago Press (1906).

Breasted, J. H., *Ancient Records of Egypt from the Earliest Times to the Persian Conquest, Collected Edited and Translated with Commentary*, Vol. III, *The Nineteenth Dynasty*, Chicago: The University of Chicago Press (1906).

Bresciani E., *Der Kampf um den Panzer des Inaros (Papyrus Krall) (MPER* Neue Serie 8), Vienna: Georg Prachner (1964).

Bresciani E., 'La corazza di Inaro era fatta con la pelle del grifone del Mar Rosso', *EVO* 13 (1990), 103–7.

Bresciani E., *Letteratura e poesia nell'antico Egitto*, 4th edn, Turin: Einaudi (2007).

Brugsch, H., *Recueil de monuments égyptiens dessinés sur lieux et publiés sous les auspices de son Altesse le Vice-Roi d'Égypte Mohammed – Saïd – Pacha, 1ere partie*, Leipzig: Hinrichs (1862).

Brugsch, H., *Thesaurus inscriptionum Aegyptiacarum: altaegyptische Inschriften (Band 5): Historisch-biographische Inschriften altaegyptischer Denkmaeler*, Leipzig: Hinrichs (1891).

Bruyère, B., *Deir el Médineh. Année 1926. Sondage au temple funéraire de Thotmès II (Hat Ankh Shesept)* (*FIFAO* IV/4), Cairo: IFAO (1952).

Bryan, B. M., *The Reign of Thutmose IV*, Baltimore and London: The Johns Hopkins University Press (1991).

Bryan, B. M., 'Pharaonic Painting through the New Kingdom' in A. B. Lloyd, ed., *A Companion to Ancient Egypt*, Chichester, UK: Wiley-Blackwell (2010).

Budge, E. A. W., *Facsimiles of Egyptian Hieratic Papyri in the British Museum with Descriptions, Summaries of Contents, etc., 2nd Series*, London: Trustees of the British Museum (1923).

Burchardt, M., *Die altkanaanäischen Fremdworte und Eigennamen in Ägyptischen*, Leipzig: Hinrichs (1910).

Burckhalter, M. and Philippa-Touchais, A., 'Salamine', *BCH* 127, livraison 2 (2003), 729–33.

Burke, A. A., 'New Light on Old Reliefs: New Kingdom Egyptian Siege Tactics and Asiatic Resistance' in J. D. Schloen, ed., *Exploring the Longue Durée: Essays in Honor of Lawrence E. Stager*, Winona Lake, IN: Eisenbrauns (2009), 57–68.

Butterlin, P., 'Mari et l'histoire militaire mésopotamienne: du temps long au temps politico-militaire' in M. d'Andrea, M. G. Micale, D. Nadali and S. Pizzamenti and A. Vacca, ed., *Pearls of the Past, Studies on Near Eastern Art and Archaeology in honour of Frances Pinnock* (marru 8), Münster: Saphron (2019), 109–39.

Calvert, A. M., 'Vehicle of the Sun: The Royal Chariot in the New Kingdom' in A. J. Veldmeijer and S. Ikram, eds, *Chariots in Ancient Egypt: The Tano Chariot, a Case of Study*, Leiden: Sidestone (2018), 45–71.

Caminos, R. A., *Late-Egyptian Miscellanies*, London: Oxford University Press (1954).

Carter, H., *The Tomb of Tut-Ankh-Amen: Discovered by the Late Earl of Carnarvon and Howard Carter*, Vol. III, London: Cassell (1933).

Carter, H. and Newberry, P. E., *The Tomb of Thoutmôsis IV*, Westminster: Archibald Constable and Co. (1904).

Càssola Guida, P., *Le armi defensive dei Micenei nelle raffigurazioni*, Rome: Edizioni dell'Ateneo (1973).

Catling, H. W., 'A Bronze Plate from a Scale-corslet found at Mycenae', *ArchAnz*, Heft 4 (1970), 441–9.

Catling, H. W., 'Panzer' in H.-G. Buchholz and J. Wiesner, *Kriegwesen, Teil 1: Schutzwaffen und Wehrbauten, Archaeologia Homerica* E 1, Göttingen: Vandenhoeck & Ruprecht (1977), 74–118.

Černý, J., *Coptic Etymological Dictionary*, Cambridge: Cambridge University Press (1976).

Cevoli, T., 'Rapporti militari tra mondo miceneo e Mediterraneo orientale alla luce di recenti scoperte archeologiche nell'isola di Salamina' in G. Borriello, *Orientalia Parthenopea* XV, Napoli: Orientalia Parthenopea Edizioni (2015), 65–78.

Champollion, J.-F., *Monuments de l'Égypte et de la Nubie d'après les dessins exécutés sur les lieux. Planches. Tome I*, Paris: Firmin Didot frères (1835).

Champollion, J.-F., *Monuments de l'Égypte et de la Nubie d'après les dessins exécutés sur les lieux. Planches. Tome III*, Paris: Firmin Didot frères (1845).

Chappaz, J.-L., 'La purification de l'or', *BSEG* 4 (1980), 19–25.

Chauveau, M., 'Les richesses méconnues de la littérature démotique', *BSFE* 156 (2003), 31–5.

Chauveau, M., 'Les romans du cycle d'Inaros et de Pédoubastis', *Egypte* 29 (2003), 19–28.

Chevereau, P.-M., *Prosopographie des Cadres Militaires Egyptiens du Nouvel Empire*, Paris: Cybèle (1994).

Cherici, A., 'Sulle rive del Mediterraneo centro-occidentale: aspetti della circolazione di armi, mercenari e culture' in G. M. Della Fina, ed., *Etruschi Greci Fenici e Cartaginesi nel Mediterraneo centrale: Atti del XIV Convegno Internazionale di Studi sulla Storia e l'Archeologia dell'Etruria* (Annali della Fondazione per il Museo Claudio Faina, Vol. XIV), Rome: Quasar (2007), 221–69.

Collombert, P. and Coulon, L., 'Les dieux contre la mer: Le début du "papyrus d'Astarté" (pBN 202)', *BIFAO* 100 (2000).

Colonna, G., 'Dischi-corazza e dischi di ornamento femminili: due distinte classi di bronzi centro-italici', *ArchClass* 58 (2007), 3–30.

Cordani, V., 'Suppiluliuma in Syria after the First Syrian War: the (Non-)Evidence of the Amarna Letters' in S. de Martino and J. L. Miller, eds, *New Results and New Questions on the Reign Suppiluliuma I*, Florence: LoGisma (2013), 43–64.

Cornelius, I., *The Iconography of the Canaanite Gods Reshef and Ba'al: Late Bronze and Iron Age I Periods (c. 1500–1000 BCE)* (*OBO* 140), Fribourg, Suisse: Editions Universitaires, Göttingen: Vandenhoeck & Ruprecht (1994).

Corteggiani, J.-P., *L'Égypte ancienne et ses dieux: Dictionnaire illustré*, Paris: Fayard (2007).

Cottevieille-Giraudet, R., *Rapport sur les Fouilles de Médamoud: Les Reliefs d'Aménophi IV Akhenaton* (*FIFAO* 13), Cairo (1936).

Couton-Perche, N., *Les armes de l'Égypte ancienne: La collection du musée du Louvre*, Paris: Édition Khéops (2021).

Cros, A. M. G., 'Mission Française de Chaldée: Campagne de 1903 Compte Rendu Sommaire des Fouilles', *RAAO* 6.1 (1904), 5–17.

Cros, A. M. G., 'Note Rectificative: Sur le Casque Chaldéen de Tello: Lettre de M. le Commandant Gaston Cros', *RAAO* 6.3 (1906), 88–9.

Crum, W. E., *A Coptic Dictionary*, Oxford: Oxford University Press (1939).

Curto, S., *L'arte militare presso gli antichi Egizi*, Turin: Edizioni d'arte Fratelli Pozzo (1973).

D'Amato, R., and Salimbeti, A., *Sea Peoples of the Bronze Age Mediterraean c. 1400 BC–1000 BC*, Oxford: Osprey Publishing (2015).

Darnell, J. C., 'Supposed Depictions of Hittites in the Amarna Period', *SAK* 18 (1991), 113–40.

Darnell, J. C. and Manassa, C., *Tutankhamun's Armies: Battle and Conquest during Ancient Egypt's Late 18th Dynasty*, Hoboken, NJ: John Wiley & Sons (2007).

Davey, C. J., 'Old Kingdom Metallurgy in Memphite Tomb Images' in L. Evans, ed., *Ancient Memphis: 'Enduring is the Perfection'. Proceedings of the International Conference held at Macquarie University, Sidney on August 14–15, 2008* (*OLA* 214), Leuven: Peeters (2012), 85–107.

Davies, W. V., *Catalogue of Egyptian Antiquities in the British Museum* VII. *Tools and Weapons* I. *Axes*, London: British Museum (1987).

Davies, W. V., 'The Tomb of Ahmose Son-of-Ibana at Elkab: Documenting the Family and other Observations' in W. Claes, H. de Meulenaere and S. Hendrickx, eds, *Elkab and Beyond: Studies in Honour of Luc Limme*, Leuven: Peeters (2009), 139–76.

Davies, W. V., 'A View from Elkab: The Tomb and Statues of Ahmose-Pennekhebet' in J. M. Galán, B. M. Bryan and P. F. Dorman, eds, *Creativity and Innovation in the Reign of Hatshepsut*, Chicago: The University of Chicago Press (2014), 381–409.

Davies, N. de Garis, 'The Graphic Work of the Expedition at Thebes', *BMMA* 23.12, Part 2: *The Egyptian Expedition 1927–1928* (Dec. 1928), 37–49.

Davies, N. de Garis, *The Tomb of Ḳen-Amūn at Thebes*, Vol. I, New York: The Metropolitan Museum of Art (1930).

Davies, N. de Garis, *The Tomb of Rekh-mi-Reʿ at Thebes*, Vol. I–II, New York: The Metropolitan Museum of Art (1943).

Davies, N. de Garis and Davies, N. de Garis, *The Tomb of Menkherrasonb, Amenmose and Another (Nos. 86, 112, 42, 226)*, London: Egypt Exploration Fund (1933).

Davies, N. de Garis and Gardiner, A. H., *Tutankhamen's Painted Box*, Oxford: Griffith Institute (1962).

Dawson, D., *The First Armies*, London: Cassell (2001).

Dawson, T., *'Armour Never Wearies': Scale and Lamellar Armour in the West, from the Bronze Age to the 19th Century*, Stroud, Gloucestershire: Spellmount (2013).

De Backer, F., 'Evolution of the Scale Armour in the Ancient Near East, Aegean and Egypt: An Overview from the Origins to the Pre-Sargonids', *ResAnt* 8 (2011), 63–104.

De Backer, F., 'Siege-Shield and Scale Armour Reciprocal Predominance and Common Evolution', *Historiae* 8 (2011), 1–29.

De Backer, F., 'Scale-Armours in the Neo-Assyrian Period: A Survey', *SAAB* XIX (2011–12), 175–202.

De Backer, F., 'Les archers de siege néo-assyriens' in G. Wilhelm, ed., *Organization, Representation, and Symbols of Power in the Ancient Near East: Proceedings of the 54th Rencontre Assyriologique Internationale at Würzburg, 20–25 July 2008*, Winona Lake, IN: Eisenbrauns (2012), 429–48.

De Backer, F., *Scale-Armours in the Neo-Assyrian Period: Manufacture and Maintenance*, Chisinau: LAP (2013).

De Backer, F., 'Cardiophylax en Urartu: un modèle celtibère' in P. Bieliński, M. Gawlikowski, R. Koliński, D. Ławecka, A. Sołtysiak and Z. Wygnańska, eds, Proceedings of the 8th International Congress on the Archaeology of the Ancient Near East: 30 April–4 May 2012, University of Warsaw. Vol. I: *Plenary Sessions,*

Township and Villages, High and Low – The Minor Arts for the Elite and for the Populace, Wiesbaden: Harrassowitz (2014), 569–84.

De Backer, F., 'Une armure expérimentale du Premier Millenaire av. J.-C.' in A. Archi, ed., Tradition and Innovation in the Ancient Near East: Proceedings of the 57th Rencontre Assyriologique Internationale at Rome 4–8 July 2011, Winona Lake, IN: Eisenbrauns (2015), 277–88.

Decker, W., 'Panzer(hemd)' in W. Helck, E. Otto, W. Westendorf, eds, Lexikon der Ägyptologie. Band 4: Megiddo – Pyramiden, Wiesbaden: Harrassowitz (1982), 665–6.

Delange, E., Monuments égyptiens du Nouvel Empire: la Chambre des Ancêtres, les Annales de Thoutmosis III et le décor de(s) palais de Séthi Ier, Paris: Louvre éditions (2015).

De Martino, S., 'The Hittite City of Šamuḫa: Its Location and Its Religious and Political Role in the Middle Kingdom' in K. Strobel, ed., New Perspectives on the Historical Geography and Topography of Anatolia in the II and I Millennium BC, Florence: LoGisma (2008), 131–43.

De Martino, S., 'The Hurrian Language in Anatolia during the Late Bronze Age' in A. Mouton, ed., Hittitology Today: Studies on Hittite and Neo-Hittite Anatolia in Honor of Emmanuel Laroche's 100th Birthday: 5e Rencontres d'archéologie de l'IFEA, Istanbul 21–22 novembre 2014, Istambul: IFEA (2017), 151–62.

De Rougé, E., 'Le Poëme de Pentaour, nouvelle tradution', RecTrav I (1870).

De Rougé, E., Oeuvres diverses, Tome V, BiEg 25, Paris: Ernest Leroux (1914).

De Rougé, J., Inscriptions hiéroglyphiques copiées en Égypte pendant la mission scientifique de M. le Vicomte Emmanuel de Rougé, Vol. IV, Paris: F. Vieweg, Libraire-Éditeur (1879).

Description de l'Égypte, ou Recueil des observations et des recherches qui ont été faites en Égypte pendant l'expédition de l'armée française, publié par les ordres de Sa Majesté l'Empereur Napoléon le Grand, Antiquités, Planches, Tome II, Paris: Imprimerie impériale (1812).

Desroches-Noblecourt, Ch., Donadoni, S. and Edel, E., Grand Temple d'Abou Simbel, II. La bataille de Qadech. Description et inscriptions. Dessins et photographies ('Coll. Scient. CEDAE' 47), Cairo: Centre de Documentation et d'Éudes sur l'Ancienne Égypte (1971).

De Vaux, R., 'Les Ḫurrites de l'Histoire et les Horites de la Bible', CRAIBL 111e année/3, 427–36.

Dezsö, T., 'Scale Armour of the 2nd Millennium BC' in T. A. Bács, ed., A Tribute to Excellence: Studies Offered in Honor of Ernő Gaál, Ulrich Luft, László Török (Studia Egyptiaca XVII), Budapest: Chaire d'Egyptologie de l'Université Eötvös Loránd (2002), 195–216.

Dezsö, T., 'Panzer' in E. Ebeling, E. Weidner, eds, *Reallexicon der Assyriologie und Vorderasiatischen Archäologie, Band* 10: Oannes – Priesterverkleidung, Berlin, New York: Walter de Gruyter (2003–5), 319–23.

Dodson, A., *Amarna Subset: Nefertiti, Tutankhamun, Ay, Horemheb and the Egyptian Counter-Reformation*, Cairo, New York: The American University in Cairo Press (2009).

Dodson, A. and Hilton, D., *The Complete Royal Families of Ancient Egypt*, Cairo: The American University in Cairo Press (2004).

Dolce, R., *Gli intarsi mesopotamici dell'epoca protodinastica*, Rome: Istituto di studi del vicino Oriente–Università (1978).

Downey, Jr., W. G., 'Captured Enemy Property: Booty of War and Seized Enemy Property', *AJIL* 44.3 (1950), 488–504.

Drenkhahn, R., *Die Handwerker und ihre Tätigkeiten im Alten Ägypten (ÄgAbh* 31), Wiesbaden: Harrassowitz (1976).

Drews, R., *The Coming of the Greeks: Indo-European Conquests in the Aegean and the Near East*, Princeton, NJ: Princeton University Press (1988).

Drews, R., *The End of the Bronze Age. Changes in Warfare and the Catastrophe ca. 1200 BC*, Princeton, NJ: Princeton University Press (1993).

Drews, R., *Early Riders: The Beginnings of Mounted Warfare in Asia and Europe*, New York, London: Routledge (2004).

Eaton-Krauss, M., 'Tutankhamun in Karnak', *MDAIK* 44 (1988), 1–11.

Edel, E., 'Die Stelen Amenophis' II aus Karnak und Memphis mit dem Bericht über den Libyerkrieg Merenptahs', *ZDPV* 69 (1953), 98–176.

Edgerton, W. F. and Wilson, J. A., *Historical Records of Ramses III: The Texts in the Medinet Habou*, Volumes I and II (*SAOC* 12), Chicago: The Oriental Institute of the University of Chicago (1936).

Eggebrecht, A., ed., *Ägyptens Aufstieg zur Weltmacht (Katalog-Handbuch zur Ausstellung im Roemer- und Pelizaeus-Museum Hildesheim 3 August–29 November 1987)*, Mainz: Philipp von Zabern (1987).

El Aguizy, O. and Haykal, F., 'Changes in Ancient Egyptian Language', *EgMa* 27–8 (1996), 25–34.

Emanuel J. P., '"Šrdn from the Sea": The Arrival, Integration, and Acculturation of a "Sea People"', *Journal of Ancient Egyptian Interconnections* 5.1 (2013), 14–27.

Erman, A. and Krebs, F., *Aus den Papyrus der Königlichen Museen (Handbücher der Staatlichen Museen zu Berlin* 8), Berlin: De Gruyter (1899).

Evelyn-White, H. G., 'The Egyptian Expedition 1914–15', *BMMA* 10 (1915), 253–6.

Faulkner, R. O., 'The Battle of Kadesh', *MDAIK* 16 (1958), 93–111.

Fischer-Elfert, W., *Die Satirische Streitschrift des P. Anastasi I. Textzusammenstellung*, Wiesbaden: Harrassowitz (1983).

Fischer-Elfert, W., *Die Satirische Streitschrift des P. Anastasi I: Übersetzung und Kommentar*, Wiesbaden: Harrassowitz (1986).

Fletcher, J., 'Ancient Egyptian Hair and Wigs', *The Ostrakon: The Journal of the Egyptian Study Society*, 13.2 (2002), 2–20.

Foster, B. R., 'The Sargonic Victory Stele from Telloh', *Iraq* 47 (1985), 15–30.

Gaballa, G. A., *Narrative in Egyptian Art*, Mainz: Philipp von Zabern (1976).

Gabolde, L., *Monuments décorés en bas relief aux noms de Thoutmosis II et Hatchepsout à Karnak* (*MIFAO* 123/1), Cairo: IFAO, 2005.

Gabolde, L. and Gabolde, M., 'Les temples "mémoriaux" de Thoutmosis II et Toutânkhamon (un rituel destine à des statues sur barques)', *BIFAO* 89 (1990), 127–78.

Gabolde, M., 'Toutânkhamon et les roseaux de Djapour' in C. Zivie-Coche and I. Guermeur, eds, *'Parcourir l'éternité': Hommages à Jean Yoyotte* (BEHE, Sciences Religieuses n. 156), Turnhout: Brepols (2012), 451–70.

Gabolde, M., 'Horemheb et les campagnes égyptiennes en Asie sous Toutânkhamon et Aÿ', *Egypte* 76 (2014/15), 19–34.

Gabolde, M., *Toutankhamon*, Paris: Pygmalion (2015).

Galán, J. M., 'Mutilation of Pharaoh's Enemy' in M. Eldamaty and M. Trad, eds, *Egyptian Museum Collections around the World: Studies for the Centennial of the Egyptian Museum, Cairo*, Vol. I, Cairo: Supreme Council of Antiquities (2002), 441–52.

Gardiner, A. H., *Egyptian Hieratic Texts. Transcribed, Translated and Annoted. Series I: Literary Texts of the New Kingdom. Part 1: The Papyrus Anastasi I and the Papyrus Koller, together with the Parallel Texts*, Leipzig: Hinrichs (1911).

Gardiner, A. H., *Late-Egyptian Stories*, Bruxelles: Édition de la Fondation égyptologique Reine Élisabeth (1932).

Gardiner, A. H., 'The *Astarte Papyrus*' in S. R. K. Glanville, ed., *Studies presented to F. Ll. Griffith*, London: Oxford University Press (1932), 74–85, pl. 8–9.

Gardiner, A. H., *Hieratic Papyri in British Museum*, Vol. I–II, 3rd Series, London: British Museum (1935).

Gardiner, A. H., *Late-Egyptian Miscellanies* (*BiAeg* 7), Brussels: Édition de la Fondation Égyptologique Reine Élisabeth (1937).

Gardiner, A. H., *Ancient Egyptian Onomastica*, Vol. 2, Oxford: Oxford University Press (1947).

Gardiner, A. H., 'The First Two Pages of the "Wörterbuch"', *JEA* 34 (1948), 12–18.

Gardiner, A. H., *The Qadesh Inscriptions of Ramesses II*, Oxford: Oxford University Press (1960).

Gardiner, A. H., *Egyptian Grammar*, 3rd edn, Oxford: Griffith Institute (2001).

Garenne-Marot, L., 'Le travail du cuivre dans l'Égypte pharaonique d'après les peintures et les bas-reliefs', *Paléorient* 11.1 (1985), 85–100.

Gernez, G., 'Des armes et des hommes: La question des modèles de diffusion des armes au Proche-Orient à l'Âge du Bronze' in P. Rouillard, ed., *Mobilités Immobilismes: L'emprunte et son refus*, Paris: De Boccard (2007), 119–34.

Gasse, A., *Données nouvelles administratives et sacerdotales sur l'organisation du domaine d'Amon. XXe–XXIe dynasties* (*BiEtud* 104), Cairo: IFAO (1988).

Gernez, G., *Les Armes du Proche-Orient des origines à 2000 av. J.-C.*, Arles: Édition Errance (2017).

Gnirs, A. M. and Loprieno, A., 'Krieg und Literatur' in R. Gundlach and C. Vogel, eds, *Militärgeschichte des pharaonischen Ägypten, Altägypten und seine Nachbarkulturen im Spiegel der aktuellen Forschung*, Paderborn, Munich, Vienna, Zürich: Schöningh (2009), 243–308.

Godlewski, W., 'Remarques sur la creation du monastère de St. Phoibammon à Deir el-Bahari', *AfrBull* 31 (1982), 107–14.

Godlewski, W., *Deir El-Bahari V: Le monastère de Phoibammon*, Warsaw: PWN–Éditions Scientifiques de Pologne (1986).

Goldwasser, O., 'What is a horse? Lexical Acculturation and Classification in Egyptian, Sumerian, and Nahuatl' in T. Pommerening and W. Bisang, eds, *Classification from Antiquity to Modern Times: Sources, Methods, and Theories from an Interdisciplinary Perspective*, Berlin, Boston: De Gruyter (2017), 45–65.

Grajetzki, W., 'Das Grab des Kii-iri in Saqqara', *JEOL* 37, 2001–2 (2003), 111–25.

Grandet, P., *Ramsés III: Histoire d'un regne*, Paris: Pygmalion (1993).

Grandet, P., *Pharaons du Nouvel Empire: une pensée stratégique (1550–1069 avant J.-C.)*, Monaco, Éditions du Rocher (2008).

Green, A. R. W., *Storm-God in the Ancient Near East*, Winona Lake, IN: Eisenbrauns (2003).

Guidotti M. C. and Pecchioli Daddi, F., eds, *La battaglia di Qadesh: Ramesse II contro gli Ittiti per la conquista della Siria*, Livorno: Sillabe (2002).

Güterbock, H. G., Hoffner, H. A., Van den Hout, T. P. J. and Goedegebuure, P. M., eds, *The Hittite Dictionary of the Oriental Institute of the University of Chicago*, Vol. Š, fasc. 2, Chicago: The Oriental Institute of the University of Chicago (2005).

Haas, V., *Geschichte der hethitischen Religion* (*HbOr* 15), Leiden, New York, Köln: Brill (1994).

Habaci, L., *The Second Stele of of Kamose and His Struggle against the Hyksos Ruler and His Capital*, Glückstadt: Verlag J. J. Augustin (1972).

Hall, R., 'The Pharaonic mss Tunic (𓌰𓏭𓂻) as a Smock', *GöttMisz* 43 (1981), 29–37.

Hallo, W. W., 'A Ugaritic Cognate for Akkadian *hitpu*?' in R. Chazan, W. W. Hallo and L. H. Schiffman, eds, *Ki Baruch Hu: Ancient Near Eastern, Biblical, and Judaic*

Studies in Honour of Baruch A. Levine, Winona Lake, IN: Eisenbrauns (1999), 78–85.

Hamblin, W. J., *Warfare in the Ancient Near East to 1600 BC: Holy Warriors at the Dawn of History*, London: Routledge (2006).

Hanson, V. D., *The Western Way of War: Infantry Battle in Classical Greece*, New York: Alfred A. Knopf (1989).

Harris, J. R., *Lexicographical Studies in Ancient Egyptian Minerals (VIO 54)*, Berlin: Akademie Verlag (1961).

Hartwig, M. K., *Tomb Painting and Identity in Ancient Thebes, 1419–1372 BCE (Monumenta Aegyptiaca 10/Imago 2)*, Turnhout: Brepol (2004).

Harvey, S., 'Monuments of Ahmose at Abydos', *EgArch* 4 (1994), 3–5.

Harvey, S., 'Abydos' in G. J. Stein, ed., *The Oriental Institute 2002–2003 Annual Report*, Chicago: The Oriental Institute of the University of Chicago (2003), 15–25.

Harvey, S., 'New Evidence at Abydos for Ahmose's Funerary Cult', *EgArch* 24 (2004), 3–6.

Hasel, M. G., *Domination and Resistance: Egyptian Military Activity in the Southern Levant, ca. 1300–1185 BC*, Leiden: Brill (1988).

Hasel, M. G., 'Israel in the Merneptah Stela', *BASOR* 296 (1994), 45–61.

Hassan, S., *Le poème dit de Pantaour et le rapport officiel sur la bataille de Qadesh*, Cairo: Imprimerie National, Égypte (1929).

Hassan, S., *The Sphinx: Its History in the Light of Recent Excavations*, Cairo: Government Press (1949).

Hassan, S., *Excavations at Gîza 8: 1936–1937: The Great Sphinx and Its Secrets: Historical Studies in the Light of Recent Excavations*, Cairo: Government Press (1953).

Healy, M., *New Kingdom Egypt* (Elite 40), Oxford: Osprey Publishing (1992).

Hayes, W. C., 'The Egyptian Expedition: The Excavation at Lisht', *BMMA* 29.11, Part 2: *The Egyptian Expedition 1933–1934* (Nov. 1934), 4–26.

Hayes, W. C., *The Scepter of Egypt: A Background for the Study of the Egyptian Antiquities in The Metropolitan Museum of Art*. Vol. II. *The Hyksos Period and the New Kingdom (1675–1080 BC)*, Cambridge, MA: The Metropolitan Museum of Art (1959).

Helck, W. *Die Beziehungen Ägyptens zu Vorderasien im 3. und 2. Jahrtausend v. Ch*, 2nd edn (*ÄgAbh* 5), Wiesbaden: Harrassowitz Verlag (1971).

Hencken, H., *The Earliest European Helmets*, Cambridge, MA: Harvard University Press (1971).

Herslund, O., 'Chariots in the Daily Life of New Kingdom Egypt: A Survey of Production, Distribution and Use in Texts' in A. J. Veldmeijer and S. Ikram, eds,

Chasing Chariots: Proceedings of the First Chariot Conference (Cairo 2012), Leiden: Sidestone (2013), 123–30.

Herslund, O., 'VI. Chronicling Chariots: Texts, Writing and Language of New Kingdom Egypt' in A. J. Veldmeijer and S. Ikram, eds, *Chariots in Ancient Egypt: The Tano Chariot, a Case of Study*, Leiden: Sidestone (2018), 150–98.

Herold, A., 'High Temperature Industries in the Late Bronze Age Capital Piramesse (Qantir): III. Workshops of The Temple of Amun Qantir/Piramesse, Site Q-I, Stratum B/2' in F. A. Esmael, ed., Z. A. Hawass, dir, *Proceedings of The First International Conference on Ancient Egyptian Mining & Metallurgy and Conservation of Metallic Artifacts, Cairo, Egypt, 10–12 April 1995*, Cairo: Ministry of Culture. Supreme Council of Antiquities (1996), 133–45.

Herold, A., 'Piramesses – The Northern Capital: Chariots, Horses and Foreign Gods' in J. Goodnick Westenholz, ed., *Capital Cities: Urban Planning and Spiritual Dimensions: Proceedings of the Symposium held on May 27–29, 1996, Jerusalem, Israel*, Jerusalem: Bible Lands Museum (1998), 129–46.

Herold, A., *Streitwagentechnologie in der Ramses-Stadt: Bronze an Pferd und Wagen*, Mainz: Philipp von Zabern (1999).

Herold, A., 'Ein Puzzle mit zehn Teilen – Waffenkammer und Werkstatt aus dem Grab des *Ky-jrj* in Saqqara' in N. Kloth K. Martin, E. Pardey, eds, *Es werde niedergelegt als Schriftstück: Festschrift für Hartwig Altenmüller zum 65. Geburtstag* (*BSAK* 9), Hamburg: Helmut Buske Verlag (2003), 193–202.

Herold, A., *Streitwagentechnologie in der Ramses-Stadt: Knäufe: Knöpfe und Scheiben aus Stain*, Mainz: Philipp von Zabern (2006).

Herold, A., 'Aspekte ägyptischer Waffentechnologie – von der Frühzeit bis zum Ende des Neuen Reiches' in R. Gundlach and C. Vogel, eds, *Militärgeschichte des pharaonischen Ägypten: Altägypten und seine Nachbarkulturen im Spiegel der aktuellen Forschung* (*Krieg und Geschichte* 34), Padeborn: F. Schöningh (2009).

Heuzey, L. and Thureau-Dagin F., 'Restitution matérielle de la Stèle des Vautours', Paris: Ernest Leroux (1909).

Hnila Gilibert, A., 'Warfare Techniques in Early Dynastic Mesopotamia', *Proceedings of the International Symposium Arms and Armour through the Ages (From the Bronze Age to the Late Antiquity), Modra-Harmónia, 19th–22nd November 2005* (*Anodos. Studies of the Ancient World* 4–5 / 2004–5), Trnava: Trnavská Univerzita (2006), 93–100.

Hoch, J. E., *Semitic Words in Egyptian Texts of the New Kingdom and Third Intermediate Period*, Princeton, NJ: University Press (2004).

Hoffmann, F., *Ägypter und Amazonen: Neubearbeitung Zweier demotischer Papyri P. Vindob D 6165 und P. Vindob D 6165 A* (*MPER* Neue Serie 24), Vienna: Hollinek (1995).

Hoffmann, F., *Der Kampf um der Panzer des Inaros: Studien zum P. Krall und seiner Stellung innerhalb des Inaros-Petubastis Zyklus (MPER* Neue Serie 26), Vienna: Hollinek (1996).

Hoffmeier, J. K., 'Reconstructing Egypt's Eastern Frontier Defense Network in the New Kingdom (Late Bronze Age)' in F. Jesse and C. Vogel, eds, *The Power of Walls – Fortifications in Ancient Northeastern Africa: Proceedings of the International Workshop held at the University of Cologne 4th–7th August 2011* (Colloquium Africanum 5), Cologne: Heinrich-Barth-Institut (2013), 163–94.

Hodgkinson, A. K., 'High-Status Industries in the Capital and Royal Cities of the New Kingdom' in A. Haducz and M. Petrik, eds, *Commerce and Economy in Ancient Egypt: Proceedings of the Third International Congress for Young Egyptologists 25–27 September 2009, Budapest* (BAR-IS 2131), Oxford: BAR Publishing (2010), 71–9.

Hodgkinson, A. K., *Technology and Urbanism in Late Bronze Egypt*, Oxford: Oxford University Press (2012).

Hoffmann, F. and Quack, J. F., *Anthologie der demotischen Literatur* (Einführungen und Quellentexte zur Ägyptologie 4), Berlin: Lit (2007).

Hoffmeier, J. K., 'Observation on the Evolving Chariot Wheel in the 18th Dynasty', *JARCE* 13 (1976), 43–5.

Hoffmeier, J. K., 'The Gebel Barkal Stela of Thutmose III (2.2 B)' in W. W. Hallo and K. L. Younger, eds, *The Context of Scripture*, Vol. 2, *Monumental Inscriptions from the Biblical World*, Leiden, Boston, Köln: Brill (2000), 14–18.

Hoffmeier, J. K., 'The Memphis and Karnak Stelae of Amenhotep II (2.3)' in W. W. Hallo and K. L. Younger, eds, *The Context of Scripture*, Vol. 2, *Monumental Inscriptions from the Biblical World*, Leiden, Boston, Köln: Brill (2000), 19–23.

Hoffmeier, J. K., 'A Possible Location in Northwest Sinai for the Sea and Land Battles between the Sea Peoples and Ramesses III', *BASOR* 380 (2018), 1–25.

Hölscher, U., *The Excavations of Medinet Habu,* Vol. II, *The Temples of the Eighteenth Dynasty* (OIP 41), Chicago: The University of Chicago Press (1939).

Homer, *The Iliad with an English translation by A. T. Murray*, Vol. 1, Cambridge, MA: Harward University Press, London: William Heinemann, Ltd. (1924).

Hovestreydt, W., 'Sideshow or not? On Side-Rooms of the First Two Corridors in the Tomb of Ramesses III' in B. J. J. Haring, O. E. Kaper and R. van Walsem, eds, *The Workman's Progress: Studies in the Village Village of Deir el-Medina and other Documents from Western Thebes in Honour of Rob Demarée*, Leiden: Peeters (2014), 103–32.

Howard, D., *Bronze Age Military Equipment*, Barnsley: Pen & Sword (2011).

Hulit, T., *Late Bronze Age scale armour in the Near East: an experimental investigation of materials, construction, and effectiveness, with a consideration of socio-economic implications*, Open Access PhD dissertation, Durham University (2002). Available at Durham E-Theses Online: http://etheses.dur.ac.uk/1006/

Ikram, S. and Dodson, A., *The Mummy in Ancient Egypt: Equipping the Dead for Eternity*, London: Thames and Hudson (1998).

Iwaszczuk, J., *Sacred Landscape of Thebes during the Reign of Hatshepsut. Royal Construction Project*. Vol. I: *Topography of the West Bank*, Warsaw: IKŚiO PAN (2017).

Iwaszczuk, J., 'Battle Scenes from the Temple of Thutmose I in Qurna', *ÄgLev* 31 (2021), 147–62.

James, T. G. H., 'Gold Technology in Ancient Egypt. Mastery of Metal Working Methods', *Gold Bulletin* 5.2 (1972), 38–42.

James, F. W. and McGovern, P. E., *The Late Bronze Egyptian Garrison at Beth Shean: A Study of Levels VII and VIII* (University Museum Monograph 85), Philadelpia: The University Museum, University of Pennsylvania (1993).

Jaroš-Deckert, B., *Grabung im Asasif 1963–1970. Band V. 'Das Grab des 'Ini-iti.f'. Die Wandmalereien der 11. Dynastie*, Mainz: Philipp von Zabern (1984).

Jenssen, J. J., *Commodity Prices from the Ramessid Period: An Economic Study of the Village of Necropolis Workmen of Thebes*, Leiden: Brill (1975).

Johnson, W. R., 'An Asiatic Battle Scene of Tutankhamun from Thebes: A Late Amarna Antecedent of the Ramesside Battle-Narrative Tradition', unpublished PhD dissertation, Department of Near Eastern Languages and Civilizations, University of Chicago (1992).

Johnson, W. R., 'Tutankhamen-period Battle Narratives at Luxor', *KMT* 20.4 (2009–10), 20–33.

Johnson, W. R. and McClain, J. B., 'A Fragmentary Scene of Ptolemy XII Worshiping the Goddess Mut and Her Divine Entourage' in S. H. D'Auria, ed., *Servant of Mut: Studies in Honor of Richard A. Fazzini*, Leiden, Boston: Brill (2008), 134–40.

Junge, F., *Late Egyptian Grammar: An Introduction*, 2nd edn (trad.: D. Warburton), Oxford: Griffith Institute (2005).

Kampp, F., *Die thebanische Nekropole: zum Wandel des Grabgedankes von der XVIII. bis zur XX. Dynastie, Teil* I, Mainz: Philipp von Zabern (1996).

Karlshausen, C. and T., 'Tutankhamen-period Battle Narratives at Luxor', *KMT* 20.4 (2009–10).

Keegan, J., *A History of Warfare*, New York: Alfred A. Knopf (1993).

Kelder, J. M., 'Royal Gift Exchange between Mycenae and Egypt: Olives as "Greeting Gifts" in the Late Bronze Age Eastern Mediterranean', *AJA* 113.3 (2009), 339–52.

Kelder, J. M., 'The Egyptian Interest in Mycenean Greece', *JEOL* 42 (2010), 125–40.

Kendall, T., 'Warfare and Military Matters in the Nuzi Tablets', unpublished PhD dissertation, Brandeis University (1974).

Kendall, T., 'gurpisu ša awēli: The Helmets of the Warriors of Nuzi' in M. A. Morrison and D. I. Owen, eds, *Studies on the Civilization and Culture of Nuzi and the Hurrians*, Vol. 1: *in Honor of Ernst R. Lacherman*, Winona Lake, IN: Eisenbrauns (1981), 201–31.

Kendall, T., 'Foreign Relations' in R. E. Freed, Y. J. Markowitz and S. H. D'Auria, eds, *Pharaohs of the Sun: Akhenaten, Nefertiti, Tutankhamen*, London: Thames & Hudson (1999), 157–61.

Kitchen, K. A., *Pharaoh Triumphant: The Life and Times of Ramesses II*, Warminster: Aris & Phillips (1982).

Klengel, H., 'Tunip und andere Probleme der historischen Geographie Mittelsyriens' in K. van Lerberghe and A. Schoors, eds, *Immigration and Emigration within the Ancient Near East. Festschrift E. Lipinski (OLA 65)*, Leuven: Peeters (1995), 125–34.

Köpp-Junk, H., 'Quellen zum Krieg im alten Ägypten' in H. Meller and M. Schefzik, eds, *Krieg – eine archäologische Spurensuche: Begleitband zur Sonderausstellung im Landesmuseum für Vorgeschichte Halle (Saale), 6. November 2015 bis 22. Mai 2016*, Halle (Saale): Landesamt für Denkmalpflege und Archäologie Sachsen-Anhalt, Landesmuseum für Vorgeschichte (2015), 229–32.

Kourou, N., 'The Aegean and the Levant in the Early Iron Age Recent Developments' in *Interconnections in the Eastern Mediterranean: Lebanon in the Bronze and Iron Ages. Proceedings of the International Symposium, Beirut 2008, BAAL* Hors-Série VI (2009), 361–74.

Krall, J., 'Der demotische Roman aus der Zeit des Königs Petubastis', *WZKM* 17 (1903), 1–36.

Krauß, R., 'Helm' in W. Helck, E. Otto, W. Westendorf, eds, *Lexikon der Ägyptologie. Band 2: Erntefest – Hordjedef*, Wiesbaden: Harrassowitz (1977), 1114–15.

Krauß, R., 'Isis' in W. Helck, E. Otto, W. Westendorf, eds, *Lexikon der Ägyptologie, Band 3: Horhekenu – Mekeb*, Wiesbaden: Harrassowitz (1980), 186–203.

Krueger, F., 'Revisiting the First Monastery of Apa Phoibammon: A Prosopography and Relative Chronology of Its Connections to the Monastery of Apa Ezekiel and the Monastic Network of Hermonthis during the 6th Century', *APF* 66.1 (2020), 150–91.

Kuentz, C., *La bataille de Qadech: les textes ('Poème de Pentaour' et 'Bulletin de Qadech') et les bas-reliefs (MIFAO 55)*, Cairo: IFAO (1928–34).

Lacherman, E. R., 'Epigraphic Evidences of the Material Culture of the Nuzians' in R. F. S. Starr, Nuzi. *Report on the Excavations at Yorgan Tepa near Kirkuk, Iraq, Conducted by Harvard University in Conjunction with the American School of Oriental Research and the University Museum of Philadelphia*, Vol. I, Cambridge, MA: Harvard University Press (1939).

Lacroix, V., 'Le "Seth asiatique" de Ramsès II, origine et justification d'un culte' in *4e édition du colloque étudiant Jean-Marie Fecteau. 27, 28, 29 mars 2019, Université du Québec à Montréal, Actes* – Vol. 2, *Revue Histoire, Idées, Sociétés,* Open access, https://revuehis.uqam.ca/colloquejeanmariefecteau/le-seth-asiatique-de-ramses-ii-origine-et-justification-dun-culte/

Lansing, A., 'Excavation at the Palace of Amenhotep III at Thebes', *BMMA* 13 Supplement (1918), 8–14.

Laroche, E., *Glossaire de la langue hurrite*, Paris: Edition Klincksieck (1980).

Lefébure, E., *Les hypogées royaux de Thèbes, seconde division: Notices des hypogées* (MMAF III, 1) Paris: Ernest Leroux (1889).

Legrain, G., 'La grande stèle d'Amenôthès II à Karnak', *ASAE* 4 (1903), 126–32.

Legrain, G., *Les temples de Karnak: fragments du dernier ouvrage de Georges Legrain*, Brussels: Vromant & Co. (1929).

Lew, D. H., 'Manchurian Booty and International Law', *AJIL* 40.3 (1946), 584–91.

Lichtheim, M., *Ancient Egyptian Literature.* Vol. II: *The New Kingdom*, Berkeley: University of California Press (1976).

Littauer, M. A. and Crouwel, J. H., 'An Egyptian Wheel in Brooklyn', *JEA* 65 (1979), 107–20.

Littauer M. A. and Crouwel, J. H., *Wheeled Vehicles and Ridden Animals in the Ancient Near East*, Leiden: Brill (1979).

Liverani, M., *Guerra e diplomazia nell'Antico Oriente 1600–1100 a.C.*, Bari: Laterza (1994).

Lolos, Y. G., 'Cypro-Mycenaean Relations ca. 1200 BC: Point Iria in the Gulf of Argos and Old Salamis in the Saronic Gulf' in N.C. Stampolidis, V. Karageorghis, eds, *ΠΛΟΕΣ. Sea Routes, Interconnections in the Mediterranean 16th–6th c. BC: Proceedings of the International Symposium held at Rethymnon, Crete, September 29th–October 2nd 2002*, Athens: The University of Crete and the A. G. Leventis Foundation (2003), 101–16.

Lolos, Y. G., 'Salamis: Kanakia' in J. Whitley, ed., *Archaeology in Greece 2003–2004* (*ArchRep* 51), London: Council the Society for the Promotion of Hellenic Studies and the British School at Athens (2004), 9–11.

Lolos, Y. G., 'Salamis: Kanakia' in J. Whitley, ed., *Archaeology in Greece 2004–2005* (*ArchRep* 50), London: Council of the Society for the Promotion of Hellenic Studies and the British School at Athens (2005), 10.

Lolos, Y. G., Marabea, C. and Oikonomou, V., 'Ajax's Capital: The Seat of the Maritime Kingdom of Salamis' in C. Pepe, ed., *Men, Lands and Seas: L'archeologia del mare. Atti del Convegno, Napoli 27–28 Giugno 2006*, Naples: Suor Orsola Benincasa (2007), 114–27.

Loprieno, A., *Ancient Egyptian: A Linguistic Introduction*, Cambridge: Cambridge University Press, 1995.

Lorenz, J. and Schrakamp, I., 'Hittite Military Warfare' in H. Genz and D. P. Mielke, eds, *Insights into Hittite History and Archaeology* (Colloquia Antiqua 2), Leuven, Paris, Walpole MA: Peeters (2011), 125-51.

Loukianoff, G., 'Un troisième texte du Poème de Pentaour sur la face oust du Temple de Louxor' *BIE* 9 (1926), 57-66.

Malek, J., 'An Early Eighteenth Dynasty Monument of Sipair from Saqqâra', *JEA* 75 (1989), 61-7.

Malnati, L., 'Armi e organizzazione militare in Etruria padana' in G. M. Della Fina, ed., *La colonizzazione etrusca in Italia: Atti del XV Convegno Internazionale di Studi sulla Storia e l'Archeologia dell'Etruria (2007)* (Annali della Fondazione per il Museo Claudio Faina, Vol. XV), Rome: Edizioni Quasar (2007), 149-86.

Manassa, C., 'Two Unpublished Memphite Relief Fragments in the Yale Art Gallery', *SAK* 30 (2002), 255-67.

Manassa, C., 'The Chariot that Plunders Foreign Lands: "The Hymn to the King in his Chariot"' in A. J. Veldmeijer and S. Ikram, eds, *Chasing Chariots: Proceedings of the First Chariot Conference* (Cairo 2012), Leiden: Sidestone (2013), 143-56.

Manniche, L., *Lost Tombs: A Study of Certain Eighteenth Dynasty Monuments in the Theban Necropolis*, London, New York: KPI (1988).

Maran, J., 'The Spreading of Objects and Ideas in the Late Bronze Age Eastern Mediterranean: Two Case Examples from the Argolid of the 13th and 12th Centuries BC', *BASOR* 336 (2004), 11-30.

Margueron, J.-C., *Mari: Métropole de l'Euphrate au IIIe millénaire av. J.-C.*, Paris: Picard (2004).

Marino, P., 'Una scena di metallurgia e oreficeria dalla tomba M.I.D.A.N.05 a Dra Abu El-Naga', *EVO* 37 (2014), 89-100.

Martin, G. T., *Corpus of Reliefs of the New Kingdom from the Memphite Necropolis and Lower Egypt*, Vol. I, London: KPI (1987).

Martin, G. T., *The Memphite Tomb of Ḥaremḥeb Commander-in-Chief of Tut'ankhamūn*, Vol. I: *The Reliefs, Inscriptions and Commentary*, London: Egypt Exploration Society, Turnhout: Brepols (1989).

Martin, G. T., *The Hidden Tomb of Memphis*, London: Thames & Hudson (1991).

Martínez Babón, J., *Historia Militar de Egipto durante la Dinastía XVIII*, Barcelona: Fundació Arqueològica Clos–Museo Egipci de Barcelona (2003).

Maspero, G., 'Le Papyrus Mallet', *RecTrav* 1 (1870), 47-59.

Maspero, G., 'Un manuel de hiérarchie égyptienne', *JournAs* 8ème 11 (1988), 250-80.

Massafra, A., *Le harpai nel Vicino Oriente antico: cronologia e distribuzione* (ROSAPAT 9), Rome: Sapienza University (2012).

Matthiae, P., 'The Victory Panel of Early Syrian Ebla: Finding, Structure, Dating', *Studia Eblaitica* 3 (2007), 33-84.

Mauric-Barberio, F., 'La tombe de Ramsès III', *Egypte* 34 (2004), 15–34.
Mauric-Barberio, F., 'Reconstitution du décor de la tombe de Ramsès III (partie inférieure) d'après les manuscrits de Robert Hay', *BIFAO* 104 (2004), 389–456.
Mazar, A., 'Beth Shean in the Second Millennium BCE: From Canaanite Town to Egyptian Stronghold' in M. Bietak, ed., *The Synchronisation of Civilization in the Eastern Mediterranean in the Second Millennium BC II: Proceedings of the SCIEM 2000 – EuroConference, Haindorff, 2nd of May–7th of May 2001*, Vienna: Verlag der Österreichischen Akademie der Wissenschafte (2003), 322–33.
Mazar, A., *Excavations at Tel Beth-Shean 1989–1996. Vol. I: From the Late Bronze Age IIB to the Medieval Period*, Jerusalem: Israel Exploration Society and the Hebrew University of Jerusalem (2006).
Mazar, A., 'Tel Beth-Shean: History and Archaeology' in R. Kratz and H. Spieckermann, eds, *One God – One Cult – One Nation: Archaeological and Biblical Perspectives*, Berlin, New York: De Gruyter (2010), 239–71.
Mazar, A., 'The Egyptian Garrison at Beth-Shean' in S. Bar, D. Kahn and J. J. Shirley, eds, *Egypt, Canaan and Israel: History, Imperialism, Ideology and Literature. Proceedings of a Conference at the University of Haifa, 3–7 May 2009*, Leiden: Brill (2011), 151–85.
McDermott, B., *Warfare in Ancient Egypt*, Gloucestershire: Sutton Publishing (2004).
McKeon, J. F. X., 'An Akkadian Victory Stele', *BMFA* 68.354 (1970), 226–43.
Minunno, G., 'Pratiche di mutilazione dei nemici caduti nel Vicino Oriente Antico', *Mesopotamia* XLIII (2008), 9–29.
Miron, R., *Kāmid el-Lōz. 10. Das 'Schatzhaus' im Palastbereich: Die Funde*, Bonn: Habelt (1990).
Mödlinger, M., 'European Bronze Age Cuirasses: Aspects of Chronology, Typology, Manufacture and Usage', *JRGZ* 59 (2012), 1–49.
Mödlinger, M., 'From Greek Boar's-Tusk Helmets to the First European Metal Helmets: New Approaches on Development and Chronology', *OJA* 32.4 (2013), 391–412.
Mödlinger, M., *Protecting the Body in War and Combat: Metal Body Armour in Bronze Age Europe* (*OREA* 6), Vienna: Austrian Academy of Sciences Press (2017).
Molleson, T. and Hodgson, D., 'The Human Remains from Woolley's Excavations at Ur', *Iraq* 65 (2003), 91–129.
Molloy, B., 'The Origins of Plate Armour in the Aegean and Europe', *Talanta* 44 (2013), 273–94.
Monnier, F., 'Proposition de reconstitution d'une tour de siège de la XIe dynastie', *JSSEA* 39 (2012–13), 125–38.
Monnier, F., 'Une iconographie égyptienne de l'architecture défensive', *ENiM* 7 (2014), 173–219.

Monnier, F., 'La scène de traction du colosse de Djéhoutyhotep: Description, traduction et reconstitution', *JAEA* 4 (2020), 55–72, Open access periodical www.egyptian-architecture.com/JAEA4/JAEA4_Monnier2.

Monnier, F., 'Les techniques de siège décrites dans la documentation pharaonique', *ENiM* 15 (2022), 51–73.

Montero Fenollós, J.-L., 'El armamento defensivo del soldado de Súmer y Mari', *AuOr* 21 (2003), 213–27.

Moorey, P. R. S., 'The Mobility of Artisans and Opportunities for Technology Transfer between Western Asia and Egypt in the Late Bronze Age' in A. J. Shortland, ed., *The Social Context of Technological Change: Egypt and the Near East, 1650–1150 BC. Proceedings of a conference held at St Edmund Hall, Oxford 12–14 September 2000*, Oxford: Oxbow (2001), 1–14.

Moran, W. L., *The Amarna Letters*, Baltimore, London: The Johns Hopkins University Press (1992).

Morentz, L. D., 'Reconsidering Sheshonk's Emblematic List and his War in Palestine' in P. Kousoulis and K. Magliveras, eds, *Moving across the Borders: Foreign Relations, Religion and Cultural Interactions in the Ancient Mediterranean* (*OLA* 159), Leuven: Peeters (2007), 101–7.

Morfini, I. and Álvarez Sosa, 'The *Min Project*. First working seasons on the unpublished Tomb of Min (TT 109) and Tomb Kampp -327-: the Tomb of May and a replica of the Tomb of Osiris' in G. Rosati and M. C. Guidotti, eds, *Proceedings of the International Congress of Egyptologist XI, Florence, Italy 23–30 August 2015*, Oxford, 2017.

Morkot, R. G., 'War and the Economy: the International "arms trade" in the Late Bronze Age and after' in T. Schneider and K. Szpakowska, eds, *Egyptian Stories: A British Egyptological Tribute to Alan B. Lloyd on the Occasion of His Retirement*, Münster: Ugarit-Verlag (2007), 169–95.

Mourad, A.-L., "Transforming Egypt into the New Kingdom: The Movement of Ideas and Technology across Geopolitical, Cultural and Social Borders' in M. Bietak and S. Prell, eds, *The Enigma of the Hyksos*, Vol. IV: *Changing Clusters and Migrations in the Near Eastern Bronze Age. Collected Papers of a Workshop held in Vienna 4th–6th of December 2019*, Wiesbaden: Harrassowitz (2021), 457–76.

Müller, H. W., *Der Waffenfund von Balâṭa-Sichem und Die Sichelschwerter*, Munich: Bayerische Akademie der Wissenschaften (1987).

Müller, M., *Akkadisch in Keilschrifttexten aus Ägypten:* Deskriptive Grammatik einer Interlanguage des späten zweiten vorchristlichen Jahrtausends anhand der Ramses-Briefe (*AOAT* 373), Münster: Ugarit–Verlag (2010).

Murnane, W. J., *The Road to Kadesh: A Historical Interpretation of the Battle Reliefs of King Sety I at Karnak* (*SAOC* 42), 2nd edn, Chicago: The University of Chicago Press (1990).

Na'aman, N., 'Yeno'am', *Tel Aviv* 4 (1977), 168–77.

Nadali, D., 'Monuments of War, War of Monuments: Some Considerations on Commemorating War in the Third Millennium BC', *Orientalia* 76, fasc. 4 (2007), 336–67.

Nadali, D., 'How Many Soldiers on the "Stele of the Vultures"? A Hypothetical Reconstruction', *Iraq* 76 (2014), 141–8.

Nagel, G., 'Set dans la barque solaire', *BIFAO* 28 (1929), 33–9.

Naville, E., *The Temple of Deir el Bahari*. Part III: *End of Northern Half and Southern Half of the Middle Platform*, London: The Egypt Exploration Fund (1898).

Naville, E., *The XIth Dynasty Temple at Deir el-Bahari*, Vol. I, London: The Egypt Exploration Fund (1907).

Nelson, H. H., 'The Naval Battle Pictured at Medinet Habu', *JNES* 2.1 (1943), 40–55.

Neveu, F., *La langue des Ramsès: Grammaire du néo-égyptien*, Paris: Éditions Khéops (1998).

Newberry, P., *El Bersheh (Band I): The Tomb of Tehuti-Hetep*, London: The Egypt Exploration Fund (1907).

Nibbi, A., 'Some Remarks on the Ancient Egyptian Shield', *ZÄS* 130 (2003), 170–81.

Nigro, L., 'La stele di Rimush da Tello e l'indicazione del rango dei vinti nel rilievo reale accadico', *Scienze dell'Antichità* 11 (2001–3), 71–93.

Obsomer, C., *Ramsès II*, Paris: Pygmalion (2012).

O'Callaghan, R. T., 'New Light on the Maryannu as "Chariot Warrior"', *JKAF* 1 (1951), 309–24.

O'Connor, 'The Sea People and the Egyptian Sources' in E. D. Oren, ed., *The Sea Peoples and Their World: A Reassessment*, Philadelphia: The University Museum, University of Pennsylvania (2000), 85–102.

Panagiotopoulos, D., 'Foreigners in Egypt in the Time of Hatshepsut and Thutmose III' in D. E. Cline and D. O'Connor, eds, *Thutmose III: A New Biography*, Ann Arbor: The University of Michigan Press (2006), 370–412.

Papadopoulos, T. J., *The Late Bronze Age Daggers of the Aegean*, Vol. 1: *The Greek Mainland* (Prähistorische Bronzefunde 6.11), Stuttgart: Franz Steiner Verlag (1998).

Papi, R., 'Produzione metallurgica e mobilità nel mondo italico' in L. Del Tutto Palma, ed., *La Tavola di Agnone nel contesto italico: Atti del convegno di studio. Agnone 13–15 aprile 1994*, Florence: Olschki Editore (1996), 89–128.

Papi, R., 'Guerrieri di pietra e dischi di bronzo', *Picus* XLI (2021), 9–84.

Parkinson, R. and Schofield, L., 'Akhenaten's Army?', *EgArch* 3 (1993), 34–5.

Parkinson, R. and Schofield, L., 'Of Helmets and Heretics: A Possible Egyptian Rappresentation of Mycenaean Warriors on a Papyrus from El-Amarna', *ABSA* 89 (1994), 157–70.

Parkinson, R. and Schofield, L., 'Images of Mycenaeans: A Recently Acquired Painted Papyrus from El-Amarna' in W. V. Davies and L. Schofield, eds, *Egypt, the Aegean and the Levant: Interconnections in the Second Millennium BC*, London: British Museum (1995), 125–6.

Parkinson, R. and Schofield, L., 'A Painted Papyrus from Amarna' in J. Phillips, ed., *Ancient Egypt, the Aegean, and the Near East: Studies in Honour of Martha Rhoads Bell*, Austin: Van Siclen Books (1998), 401–6.

Parrot, A., 'Les fouilles de Mari: Prèmiere campagne (hiver 1933–1934)', *Syria* 16, fasc. 2 (1935), 117–40.

Parrot, A., *Tello: Vingt campagnes de fouilles (1877–1933)*, Paris: Albin Michel (1948).

Parrot, A., *Mission archéologique de Mari*, Vol. I, *le temple d'Ishtar* (Insitut fr. d'Archéologie de Beyrouth, t. LXV), Paris: Paul Geuthner (1956).

Parrot, A., 'Les fouilles de Mari: Dix-neuvième campagne (printemps 1971)', *Syria* 48, fasc. 3–4 (1971), 253–70.

Partridge, R. B., *Fighting Pharaohs: Weapons and Warfare in Ancient Egypt*, Manchester: Peartree Publishing (2002).

Pendlebury, J. D. S., *The City of Akhenaten. Part III: The Central City and the Official Quarters: The Excavations at Tell el-Amarna during the Seasons 1926–1927 and 1931–1936. Volume One: Text*, London (1951).

Petschel, S. and von Falck, M., eds, *Pharao siegt immer: Krieg und Frieden im Alten Ägypten*, Hamm: Kettler (2004).

Petersen, L. and Kehrer, N., ed., *Ramses: Göttlicher Herrscher am Nil*, Karlsruhe: Michael Imhof Verlag (2016).

Peterson, B. E. J., *Zeichnungen aus einer Totenstadt: Bildostraka aus Theben-West, ihre Fundplätze, Themata und Zweckbereiche litsamt einem Katalog der Gayer-Anderson-Sammlung in Stockholm (BMMNEA 7-8)*, Stockholm (1973).

Petrie, W. M. P., *Tools and Weapons Illustrated by the Egyptian Collection in University College, London, and 2,000 Outlines from other Sources*, London: British School of Archaeology in Egypt and Egyptian Research (1917).

Peust, C., *Egyptian Phonology. An Introduction to the Phonology of a Dead Language*, Göttingen: Peust & Gutschmidt Verlag (1999).

Philippa-Touchais, A., 'Salamine', *BCH* 128–9, livraison 2.2 (2004), 1299–300.

Pietri, R., 'Le roi en char au Nouvel Empire', *Egypte* 74 (2014), 13–22.

Pinch, G., *Magic in Ancient Egypt*, London: British Museum (1994).

Pollastrini, A. M., 'La poliorcetica in Egitto dall'Antico Regno alla XXV dinastia' in P. Gallo, ed., *Egittologia a Palazzo Nuovo: Studi e ricerche dell'Università di Torino*, Novi Ligure: Edizioni Epoké (2013), 237–58.

Pollastrini, A. M., 'Una rappresentazione 'tridimensionale' di tarda Età del Bronzo?' in A. Di Natale and C. Basile, eds, *Atti del XVIII Convegno di Egittologia e*

Papirologia, Siracusa, 20–23 Settembre 2018 (*Quaderni del Museo del Papiro* XVII), Siracuse: Tyche edizioni (2020), 143–51.

Pollastrini, A. M., 'Le armi dei popoli vinti nell'iconografia egiziana' in S. Graziani and G. Lacerenza, eds, *Egitto e Vicino Oriente antico tra passato e future. The Stream of Tradition: la genesi e il perpetuarsi delle tradizioni in Egitto e nel Vicino Oriente antico*, Naples: UniorPress (2022), 305–15.

Prell, S., 'A Glimpse into the Workshops of the Chariotry of Qantir-Piramesse – Stone and Metal Tools of Site Q I' in A. J. Veldmeijer and S. Ikram, eds, *Chasing Chariots: Proceedings of the First Chariot Conference* (Cairo 2012), Leiden: Sidestone (2013), 157–74.

Prisse d'Avennes, A., *Histoire de l'art égyptien: d'après les monuments; depuis les temps les plus reculés jusqu'à la domination romaine, Atlas, Tome II*, Paris: Arthus Betrand (1879).

Prisse d'Avennes, E., *Monuments égyptiens, bas-reliefs, peintures, inscriptions, etc., d'après les dessins exécutés sur les lieux par E. Prisse d'Avennes, pour faire suite aux Monuments de l'Égypte et de la Nubie de Champollion-le-Jeune*, Paris: Firmin Didot frères (1847).

Pritchard, J. B., 'Syrians as Pictured in the Paintings of the Theban Tombs', *BASOR* 122 (1951), 36–41.

Pusch, E. B., 'Metallverarbeitende Werkstätten der frühen Ramessidenzeit in Qantir-Piramesse/Nord', *ÄgLev* 1 (1990), 75–113.

Pusch, E. B., '"Pi-Ramesse-geliebt-von-Amun, Hauptquartier Deiner Streitwagentruppen": Ägypter un Hethiter in der Delta-Reziden der Ramessiden' in A. Eggebrecht, ed., *Pelizaeus-Museum Hildesheim: Die ägyptische Sammlung*, Mainz: Philipp von Zabern (1993), 126–43.

Pusch, E. B., 'Divergierende Verfahren der Metallverarbeitung in Theben und Qantir? Bemerkungen zu Konstruktion und Technik', *ÄgLev* 4 (1994), 145–70.

Pusch, E. B., 'High Temperature Industries in the Late Bronze Age Capital Piramesse (Qantir): II. A Quasi-industrial Bronze Factory Installation, Tools and Artifacts' in F. A. Esmael, ed., Z. A. Hawass, dir., *Proceedings of The First International Conference on Ancient Egyptian Mining & Metallurgy and Conservation of Metallic Artifacts, Cairo, Egypt, 10–12 April 1995*, Cairo: Ministry of Culture, Supreme Council of Antiquities (1996), 121–32.

Pusch, E. B., 'Vorbericht über die Abschlusskampagne am Grabungsplatz Q IV 1997', *ÄgLev* 9 (1999), 17–37.

Pusch, E. B., 'Piramesse-Qantir. Residenz, Waffenschmiede und Drehscheibe internatinaler Beziehungen' in S. Petschel and M. von Falck, *Pharao siegt immer: Krieg und Frieden im alten Ägypten*, Bönen: Kettler (2004), 239–63.

Pusch, E. B. and Rehen, T., *Hochtemperatur-Technologie in der Ramses-Stadt: Rubinglas für den Pharao*, Vol. 1–2 (*Forschungen in der Ramses-Stadt* 6), Hildesheim: Gerstenberg (2007).

Quibell, J. E., *Excavations at Saqqara (1908–9, 1909–10): The Monastery of Apa Jeremias*, Cairo: Imprimerie IFAO (1912).

Quibell, J. E. and Hayter A. G. K., *Excavations at Saqqara: Teti Pyramid, North Side*, Cairo: Imprimerie IFAO (1927).

Rademakers, F. W., Rehren, T. and E. B. Pusch, 'Bronze Production in Pi-Ramesse: Alloying Technology and Material Use' in E. Ben-Yosef, ed., *Mining for Ancient Copper: Essays in Memory of Beno Rothenberg* (Monograph Series of the Sonia and Marco Nadler Institute of Archaeology 37) University Park, PA/Tel Aviv: Eisenbrauns & Emery and Claire Yass Publications in Archaeology of the Institute of Archaeology, Tel Aviv University (2018), 503–25.

Radwan, A., 'The First Appearance of Isis in a Foreign Dress', *Memnonia* IX (1998), 175–80.

Ragazzoli, C., 'Les manuscripts de miscellanies en Égypte ancienne, ou la lecture comme pratique créative' in S. Morlet, ed., *Lire en extraits: Une contribution à l'histoire de la lecture et de la literature, de l'Antiquité au Moyen Âge*, Paris: Presses de l'Université Paris-Sorbonne (2015), 11–28.

Randall-Maciver, D. and Woolley, C. L., *The Eckley B. Coxe Junior Expedition to Nubia*: Vol. VII, *Buhen (text)*, Philadelphia: The Universty Museum (1911).

Rainey, A. F., 'The Soldier-Scribe in Papyrus Anastasi I', *JNES* 26 (1967), 58–60.

Raulwing, P., 'The Kikkuli Text (CTH 284): Some Interdisciplinary Remarks on Hittite Training Texts for Chariot Horses in the Second Half of the 2nd Millennium BC' in A. Gardeisen, ed., *Les Équidés dans le monde méditerranéen antique. Actes du colloque organise par l'École française d'Athènes, le Centre Camille Jullian et l'UMR 5140 du CNRS, Athèns, 26–28 novembre 2003*, Lattes: Édition de l'Association pour le développement de l'archéologie en Languedoc-Rousillon (2005), 61–75.

Redford, D. B., 'Foreigners (Especially Asiatics) in the Talatat' in D. Redford, ed., *The Akhenaten Temple Project*, Vol. 2: *Rwd – mnw, Foreigners and Inscriptions*, Toronto: University of Toronto Press (1988), 13–27.

Redford, D. B., *Egypt, Canaan and Israel in Ancient Times*, Princeton: Princeton University Press (1992).

Redford, D. B., *The Wars in Syria and Palestine of Thutmose III*, Leiden, Boston: Brill (2003).

Redford, D. B., 'The Northern Wars of Thutmose III' in D. E. Cline and D. O'Connor, eds, *Thutmose III: A New Biography*, Ann Arbor: The University of Michigan Press (2006), 325–43.

Reisner, G. A., 'Inscribed Monuments from Gebel Barkal', *ZÄS* 66 (1931), 76–100.
Reisner, G. A. and Reisner M. B., 'Inscribed Monuments from Gebel Barkal, Part 2: The Granite Stela of Thutmose III', *ZÄS* 69 (1933), 24–39.
Reviv, H., 'Some Comments on the Maryannu', *IEJ* 22 (1972), 218–28.
Richter, T., *Bibliographisches Glossar des Hurritischen*, Wiesbaden: Harrassowitz (2012).
Robinson, H. R., *Oriental Armour*, London: Jenkins (1967).
Roccati, A., 'Il bilinguismo interno dell'Egitto', *VicOr* 3 (1980), 77–84.
Roccati, A., 'La lingua diffusa (politica e lingua nell'Egitto ramesside)', *ParPass* 268 (1993).
Romano, L., 'La Stele degli Avvoltoi: Una Rilettura Critica', *VicOr* XIII (2007), 3–23.
Rommelaere, C., *Les chevaux du Nouvel Empire égyptien: Origines, races, harnacement*, Brussels: Safran (1991).
Rosellini, I., *I Monumenti Storici dell'Egitto e della Nubia. Tomo I: Monumenti Storici*, Pisa: Capurro (1832).
Rosellini, I., *I Monumenti Storici dell'Egitto e della Nubia. Tomo II: Monumenti Civili*, Pisa: Capurro (1834).
Rostoker, W., 'Ancient Techniques for Making Holes in Sheet Metal', *AJA* 90 (1986), 93–4.
Ryholt, K., 'The Assyrian Invasion of Egypt in Egyptian Literary Tradition' in J. G. Derckson, ed., *Assyria and Beyond: Studies Presented to Mogens Trolle Larsen*, Leiden: Nederlands Instituut voor het Nabije Oosten (2004), 483–510.
Sa'ad, R., 'Fragments d'un monument de Toutânkhamon retrouvés dans le IXe pylône de Karnak', *CahKarn* 5 (1975), 93–109.
Sabbahy, L., 'V. Moving Pictures: Context of Use and Iconography of Chariots in the New Kingdom' in A. J. Veldmeijer and S. Ikram, eds, *Chariots in Ancient Egypt: The Tano Chariot, a Case of Study*, Leiden: Sidestone (2018), 120–49.
Sanchez, G. M., 'A Neurosurgeon's View of the Battle of Reliefs of King Sety I: Aspects of Neurological Importance', *JARCE* 37 (2000), 143–65.
Sandars, N. K., *The Sea Peoples: Warriors of the Ancient Mediterranean. 1250–1150 BC*, Revised edn, London: Thames and Hudson (1985).
Sauneron, S., 'La manufacture d'armes de Memphis', *BIFAO* 54 (1954), 7–12.
Sayce, A. H., 'The Astarte Papyrus and the Legend of the Sea', *JEA* 19 (1933), 56–9.
Schaden, O. J., 'Tutankhamun and Ay Blocks from Karnak', *NARCE* 80 (1972), 39–40.
Schaden. O. J., 'A Tutankhamun Stela at Karnak', *CahKarn* 8 (1978), 279–84.
Schaden, O. J., 'Tutankhamon-Ay Shrine at Karnak and Western Valley of the Kings Project. Report on the 1985–1986 Season', *NARCE* 138 (1978), 10–15.
Schaden, O. J., 'Clearance of the Tomb of Ay (WV-23)', *JARCE* 21 (1984), 39–64.
Schaden, O. J., 'Report on the 1978 Season at Karnak', *NARCE* 27 (1984), 44–64.

Scheel, B., 'Studien zum Metallhandwerk in Alten Ägypten I: Handlungen und Beischriften in den Bildprogrammen der Gräber des Alten Reiches', *SAK* 12 (1985), 117–77.

Scheel, B., 'Studien zum Metallhandwerk in Alten Ägypten II: Handlungen und Beischriften in den Bildprogrammen der Gräber des Mittleren Reiches', *SAK* 13 (1986), 181–205.

Scheel, B., 'Studien zum Metallhandwerk in Alten Ägypten III: Handlungen und Beischriften in den Bildprogrammen der Gräber des Neuen Reiches und der Spätzeit', *SAK* 14 (1987), 247–64.

Scheel, B., *Egyptian Metalworking and Tools* (ShirEgypt 13), Buckinghamshire: Shire (1989).

Schneider, T., 'Foreign Egypt: Egyptology and the Concept of Cultural Appropriation', *ÄgLev* 13 (2003), 155–60.

Schneider, T., 'Fremdwörter in der ägyptischen Militärsprache der Neuen Reiches und ein Bravourstück des Elitesoldaten (Papyrus Anastasi I 23, 2–7)', *JSSEA* 35 (2008), 181–205.

Schrakamp, I., *Krieger und Waffen im frühen Mesopotamien: Organisation und Bewaffnung des Militärs in frühdynastischer und sargonischer Zeit*, Magdeburg: Philipps-Universität (2010).

Schulman, A. R., 'Egyptian Representations of Horsemen and Riding in the New Kingdom', *JNES* 16 (1957), 263–71.

Schulman, A. R., 'The Egyptian Chariotry: A Reexamination', *JARCE* 2 (1963), 75–98.

Schulman, A. R., *Military Rank, Title and Organization in the Egyptian New Kingdom* (*MÄS* 6), Berlin: Verlag Bruno Hessling (1964).

Schulman, A. R., 'Some Observations on the Military Background of the Amarna Period', *JARCE* 3 (1964), 51–69.

Schulman, A. R., 'Hittite, Helmets and Amarna: Akhenaten's First Hittite War' in D. Redford, ed., *The Akhenaten Temple Project*, Vol. 2: *Rwd – mnw, Foreigners and Inscriptions*, Toronto: University of Toronto Press (1988), 53–79.

Sethe, K., *Untersuchungen zur Geschichte und Altertumskunde Aegyptens*, Leipzig: Hinrichs (1896).

Sethe, K., 'Mißverstandene Inschriften', *ZÄS* 44 (1907), 35–41.

Simon, C., 'Les campagnes militaires de Ramsès III à Médinet Habou: Entre vérité et propagande' in C. Karlshausen and C. Obsomer, eds, *De la Nubie à Qadech: La guerre dans l'Égypte ancienne*, Brussels: Édition Safran (2016), 171–94.

Singer, I., 'Merneptah's Campaign to Canaan and the Egyptian Occupation of the Southern Costal Plain of Palestine in the Ramesside Period', *BASOR* 269 (1988), 1–10.

Smith, H. S., *The Fortress of Buhen: The Inscriptions*, London: Egypt Exploration Society (1976).

Smith, W. S., *Interconnections in the Ancient Near East: A Study of the Relationships between the Arts of Egypt, the Aegean, and Western Asia*, New Haven, CN, London: Yale University Press (1965).

Snodgrass, A. M., *The Arms and Armour of the Greeks,* Ithaca, New York: Cornell University Press (1967).

Spalinger, A. J., 'A Critical Analysis of the "Annals" of Thutmose III (Stücke V–VI)', *JARCE* XIV (1977), 41–54.

Spalinger, A. J., 'A Canaanite Ritual Found in Egyptian Reliefs', *JSSEA* 8 (1978), 47–60.

Spalinger, A. J., 'The Northern War of Seti I: An Integrative Study', *JARCE* 16 (1979), 29–47.

Spalinger, A. J., 'Historical Observations on the Military Reliefs of Abu Simbel and other Ramesside Temples in Nubia', *JEA* 66 (1980), 83–99.

Spalinger, A. J., *The Transformation of an Ancient Egyptian Narrative: P. Sallier III and the Battle of Kadesh* (*GOF* IV/40), Wiesbaden: Harrassowiz (2002).

Spalinger, A. J., *War in Ancient Egypt: The New Kingdom*, Oxford: Blackwell (2005).

Spalinger, A. J., *Icons of Power: A Strategy of Reinterpretation*, Prague: Charles University in Prague, Faculty of Arts (2011).

Spalinger, A. J., 'Re-Reading Egyptian Military Reliefs' in M. Collier and S. Snape, eds, *Ramesside Studies in Honour of A. K. Kitchen*, Bolton: Rutherford Press (2011), 475–91.

Spalinger, A. J., 'Egyptian Chariots: Departing for War' in A. J. Veldmeijer and S. Ikram, eds, *Chasing Chariots: Proceedings of the First Chariot Conference* (Cairo 2012), Leiden: Sidestone (2013), 237–56.

Spalinger, A. J., *Leadership under Fire: The Pressure of Warfare in Ancient Egypt*, Paris: Éditions Soleb (2020).

Spalinger, A. J., 'Le positionnement des troupes sur le champs de bataille du Nouvel Empire', *Egypte* 106 (2022), 3–16.

Speiser, E. A., 'On Some Articles of Armor and Their Names', *JAOS* 70.1 (1950), 47–9.

Spiegelberg, W., *Hieratic Ostraka and Papyri Found by J. E. Quibell in the Ramesseum*, London: Quaritch (1898).

Spiegelberg, W., *Der Sagenkreis des Königs Petubastis nach dem Strassburger demotischen Papyrus sowie den Wiener und Pariser Bruchstücken* (*DemStud* 3), Leipzig: Hinrichs (1910).

Spieser, C., *Les noms du Pharaon comme êtres autonomes au Nouvel Empire* (*OBO* 174), Fribourg, Suisse: Editions Universitaires, Göttingen: Vandenhoeck & Ruprecht (2000).

Spieser, C., 'Cartouche' in E. Frood and W. Wendrich, eds, *UCLA Encyclopedia of Egyptology*, Los Angeles (2010), 1–9, https://escholarship.org/uc/item/3g726122.

Stadelmann, R., *Syrisch – Palästinensische Gottheiten in Ägypten*, Leiden: Brill (1967).

Stager, L. E., 'Merenptah, Israel and Sea Peoples: New Light on an Old Relief', *ErIsr* 18 (1985), 56–64.

Staring, N., *The Saqqara Necropolis through New Kingdom: Biography of an Ancient Egyptian Cultural Landscape* (Culture and History of the Ancient Near East, Vol. 131), Leiden: Brill (2022).

Starr, R. F. S., *Nuzi: Report on the Excavations at Yorgan Tepa near Kirkuk, Iraq, conducted by Harvard University in conjunction with the American School of Oriental Research and the University Museum of Philadelphia*, Vol. I, Cambridge, MA: Harvard University Press (1939).

Tallet, P., *Sésostris III et la fin de la XIIe dynastie*, Paris: Pygmalion (2015).

te Velde, H., *Seth, God of Confusion: A Study of His Role in Egyptian Mythology and Religion* (*Probl Äg* 6), Leiden: Brill (1967).

The Epigraphic Survey, *Medinet Habu*, Vol. I, *Earlier Historical Records of Ramses III* (*OIP* 8), Chicago: The University of Chicago Press (1930).

The Epigraphic Survey, *Medinet Habu*, Vol. II, *Later Historical Records of Ramses III* (*OIP* 9), Chicago: The University of Chicago Press (1939).

The Epigraphic Survey, *The Temple of Khonsu*, Vol. I, *Scenes of King Herihor in the Court with Translation of Texts* (*OIP* 100), Chicago: The University of Chicago Press (1979).

The Epigraphic Survey, *Relief anf Inscriptions at Karnak*, Vol. IV, *The Battle Reliefs of King Sety* I (*OIP* 107), Chicago: The University of Chicago Press, 1986.

Thomas, A. and Potts, T., *Mesopotamia: Civilization Begins*, Los Angeles: Getty Publications (2020).

Thordeman, B., 'The Asiatic Splint Armour in Europe', *AcArch (C)* 4 (1933), 117–50.

Thordeman, B., *Armour from the Battle of Wisby 1361* (2 vols), Stockholm: Kungl. Vitterhets, historie och antikvitets akademien (1936).

Tomedi, G., *Italische Panzerplatten und Panzerscheiben: Prähistorische Bronze-funde, Abteilung* III, *Band* 3, Stuttgart: Franz Steiner Verlag (2000).

Trémouille, M. C., 'Šauška, Šawuška' in E. Ebeling, E. Weidner, *Reallexicon der Assyriologie und Vorderasiatischen Archäologie, Band* 12: 'Šamuḫa–Schild', New York: Walter de Gruyter (2009), 99–103.

Trimm, C., *Fighting for the King and the Gods: A Survey of Warfare in the Ancient Near East*, Atlanta: SBL Press (2017).

Turner, S., *The Horse in the New Kingdom Egypt: Its Introduction, Nature, Role and Impact*, Wallasey: Abercromby Press (2021).

van Soldt, W. H., 'Akkadian as a Diplomatic Language' in S. Weninger, ed., *The Semitic Languages: An International Handbook*, Berlin: De Gruyter Mouton (2011), 405–15.

van Soldt, W. H., 'The Orontes Valley in texts from Alalaḫ and Ugarit during the Late Bronze Age, ca. 1500–1200 BC', *Syria* IV (2016), 137–44.

Vanschoonwinkel, J., 'Les Peuples de la Mer d'après une lecture archéologique des reliefs de Médinet Habou' in C. Karlshausen and C. Obsomer, eds, *De la Nubie à Qadech: La guerre dans l'Égypte ancienne*, Brussels (2016), 193–234.

Veldmeijer, A. J., *Amarna's Leatherwork. Part I: Preliminary Analysis and Catalogue*, Leiden: Sidestone (2011).

Veldmeijer, A. J. and Ikram, S., eds, *Chasing Chariots: Proceedings of the First Chariot Conference* (Cairo 2012), Leiden: Sidestone (2013).

Veldmeijer, A. J. and Ikram S., eds, *Chariots in Ancient Egypt: The Tano Chariot, a Case of Study*, Leiden: Sidestone (2018).

Veldmeijer, A. J., Hulit, T., Skinner, L.-A. and Ikram, S., 'Tutankhamun's Cuirass Reconsidered', *JEOL* 48 (2021-2), 125–56.

Ventzke, W., 'Zur Rekonstruktion eines bronzenen Schuppenpanzer' in R. Hachmann, ed., *Frühe Phöniker im Libanon: 20 Jahre deutsche Ausgrabungen in Kāmid el-Loz*, Maiz: Philipp von Zabern (1983), 94–100.

Vernus, P., 'Égyptian', *École pratique des hautes études. 4e section, sciences historiques et philologiques. Livret 4. Rapports sur les conférences des années 1985–1986 & 1987*, Paris: École pratique des hautes études (1994).

Vernus, P., 'Réception linguistique et idéologique d'une nouvelle technologie: le cheval dans la civilisation pharaonique' in M. Wissa, ed., *The knowledge economy and technological capabilities of Egypt, the Near East and the Mediterranean, 2nd millennium BC–1st millennium AD: proceedings of a confererence held at the Maison de la Chimie, Paris, France 9–10 december 2005*, Sabadell (Barcelona): Ausa (2010), 1–46.

Vikentiev, V., 'Les rites de la réinvestiture royale en tant que champ de recherches sur la période archaïque égypto-libyenne', *BIE* 37 (1957), 271–316, pl. I–V.

Virey, Ph., 'Le tombeau d'un Seigneur de Thini dans la nécropole de Thèbes', *RecTrav* 9 (1887), 27–32.

Virey, Ph., *Le tombeau de Rekhmara, Préfet de Thèbes sous la XVIIIe dynastie* (*MMAF* V, 1), Paris: Ernest Leroux (1889).

Virey, Ph., *Sept Tombeaux thébains de la XVIIIe dynastie* (*MMAF* V, 2), Paris: Ernest Leroux (1891).

Vogel, C., 'Fallen Heroes? – Winlock's "Slain Soldiers" Reconsidered', *JEA* 89 (2003), 239–45.

Vogel, C., *Ägyptische Festungen und Garnisonen bis zum Ende des Mittleren Reiches* (*HÄB* 46), Hildesheim: Gerstenberg (2004).

Vogel, C., 'Hieb- und stichfest? Überlegungen zur Typologie des Sichelschwertes im Neuen Reich' in D. Bröckelmann and A. Klug, eds, *In Pharaos Staat: Festschrift für Rolf Gundlach zum 75. Geburtstag*, Wiesbaden: Harrassowitz (2006), 271–86.

Vogelsang-Eastwood, G., *Pharaonic Egyptian Clothing*, Leiden, New York, Köln: Brill (1993).

Volokine, Y., *La Frontalité dans l'iconographie de l'Égypte ancienne* (*CSEG* 6), Geneva: Société d' Égyptologie (2000).

Volten, A., *Ägypter und Amazonen: Eine demotische Erzählung des Inaros-Petubastis-Kreises aus zwei Papyri der Österreichen Nationalbibliothek (Pap. Dem. Vindob. 6165 und 6165 A)* (MPER Neue Serie 6), Vienna: Hollinek (1962).

Von Beckerath, J., *Handbuch der ägyptischen Königsnamen* (*MÄS* 49), Mainz: Philipp von Zabern (1988).

Vycichl, W., *Dictionnaire étimologique de la langue copte*, Leuven: Peeters (1983).

Warmenbol, E., *Sphinx: les gardiens de l'Égypte*, Brussels: Mercator (2006).

Wegner, I., *Gestalt und Kult der Ištar-Šawška in Kleinasien* (*AOAT* 36) Neukirchen – Vluyn: Butzon und Bercker (1981).

Wegner, I., *Corpus der hurritischen Sprachdenkmäler (ChS) I. Abteilung: Die Texte aus Boğazköy, Band 3. Hurritische Opferlisten aus hithitischen Festbeschreibungen, Teil. 1 Texte für Ištar – Ša(w)uška*, Rome: Bonsignori (1995).

Wiedemann, A., *Hieratische Texte aus den Museen zu Berlin und Paris*, Leipzig: J. A. Barth (1879).

Wiese, A., Winterhalter, S. and Brodbeck, A. (coll.), *Antikenmuseum Basel und Sammlung Ludwig: Die Ägyptische Abteilung*, Mainz: Philipp von Zabern (2001).

Wilhelm, G., *The Hurrians* (trad.: J. Barnes), Warminster: Aris & Phillips (1989).

Wilkinson, J. G., *The Manner and Customs of the Ancient Egyptians*. Vol. I, London: Dodd, Mead & Co. (1878).

Willems, H. O., 'The Nomarchs of the Hare Nome and Early Middle Kingdom History', *JEOL* 28 (1983–4), 80–102.

Wilson, J. A., 'Egyptian Historical Texts' in J. B. Pritchard, ed., *ANET*, Princeton, NJ: Princeton University Press (1954), 227–64.

Winand, J., 'Identifying Semitic Loanwords in the Late Egyptian' in E. Grossman, P. Dils, T. S. Richter and W. Schenkel, eds, *Greek Influence on Egyptian-Coptic: Contact-Induced Change in an Ancient African Language*, Hamburg: Widmaier (2017), 481–511.

Winlock, H. E., 'The Work of The Egyptian Expedition', *BMMA* 7 (1912), 184–90.

Winlock, H. E., 'Excavations at Thebes, 1919–20', *BMMA* 15 Supplement (1920), 12–32.

Winlock, H. E., 'The Egyptian Exploration, 1925-1927: The Museum's Excavations at Thebes', *BMMA* 23 (1928), 3-58.
Winlock, H. E., *Excavations at Deir el Baḥri 1911-1912*, New York: The Macmillan Company (1942).
Winlock, H. E., 'The Eleventh Dynasty', *JNES* 2 (1943), 249-83.
Winlock, H. E., *The Slain Soldiers of Neb-ḥetep-Rē' Montu-ḥotpe* (MMA Exp. 16), New York: Egyptian Expedition Publications (1945).
Winlock, H. E., *The Rise and Fall of the Middle Kingdom in Thebes*, New York: The Macmillan Company (1947).
Winter, I. J., 'After the Battle Is Over: The "Stele of the Vultures" and the Beginning of Historical Narrative in the Art of the Ancient Near East', *Symposium Papers IV: Pictorial Narrative in Antiquity and the Middle Ages* (*Studies in the History of Art*, Vol. 16), Washington: National Gallery of Art, 1985, 11-35.
Wise, T., *Ancient Armies of the Middle East* (Men-at-Arms 109), Oxford: Osprey Publishing (1981).
Wolf, W., *Die Bewaffnung des altägyptischen Heeres*, Leipzig: Hinrichs (1926).
Woolley, C. L., *The Royal Cemetery: A Report on the Predynastic and Sargonid Graves Excavated between 1926 and 1931* (2 vols), Ur Excavations, Vol. 2, Oxford: Oxford University Press (1934).
Wreszinski, W., *Atlas zur altägyptischen Kulturgeschichte. II. Teil*, Leipzig: Hinrichs (1935).
Yadin, Y., *The Art of Warfare in Biblical Lands in the Light of Archaeological Discovery* (2 vols), New York: McGraw-Hill (1963).
Yadin, Y., 'The Earliest Representation of a Siege Scene and a "Scythian Bow" from Mari', *IEJ* 22 (1972), 89-94.
Yeivin, Sh., 'Amenophis II's Asiatic Campaigns', *JARCE* 6 (1967), 119-28.
Youssef, A. A.-H., 'A Nineteenth Dynasty New Word for Blade and the Semitic Origin of Some Egyptian Weapon-Names and Other Related Words', *MDAIK* 39 (1983), 255-60.
Youssef, A. A.-H., Leblanc, Ch. and Maher, M., *Ramesséum IV: Les batailles de Tounip et de Dapour* ('Coll. Scient. CEDAE'), Cairo (1977).
Yoyotte, J., 'Les Stèles de Ramsès II a Tanis: Première Partie', *Kêmi* 10 (1949), 58-74.
Yoyotte, J. and López, J., 'L'organisation de l'armée et les titulatures des soldats au nouvel empire égyptien', *BiOr* 26.1-2 (1969), 3-19.
Yurco, F. J., 'Meremptah's Canaanite Campaign', *JARCE* 22 (1986), 189-215.
Yurko, F. J., '3,200-Year-Old Picture of Israelites Found in Egypt', *BAR* 16.5 (September/October 1990), 20-38.
Zandee, J., 'Seth als Sturmgott', *ZÄS* 90 (1963), 144-56.

Zeidler, J., 'A New Approach to the Late Egyptian "Syllabic Orthography"' in *Sesto Congresso Internazionale di Egittologia, Atti*, Vol. II, Turin: Tipografia Torinese – Stabilimento Poligrafico (1993).

Zivie, C. M., *Giza au deuxième millénaire* (*BiEtud* 70), Cairo: IFAO (1976).

Zivie-Coche, C. M., *Giza au premier millénaire: Autour du temple d'Isis Dame des Pyramides*, Boston (1991).

Zorn, J. R., 'Reconsidering Goliath: An Iron Age I Philistine Chariot Warrior', *BASOR* 360 (2010), 1–22.

Index

Figures are denoted by the use of *italics*. Notes are labelled with 'n.'

Abu Simbel 32, *33*, 45, *46*, 54, 57, 96, 127 n.61
Ahab of Israel 124 n.34
Ahmose I 23–4, 122 n.23
Ahmose, son of Ibana 21, 120 n.3
Akhenaten 27
Albright, William 97
Amarna 6, 27, 41, 105
Amenemhat II 16, 95
Amenhotep I 122 n.23
Amenhotep II 37–8, *39*, 40–1, 58, 94–5
Amenhotep III 69–70, 93, 129 n.82, 136 n.33
Amenhotep IV *see* Akhenaten
Amurru 32, *33*
armour 15–16, 17, 22, *23*, 27, 40, 41, 43, 45, *46*, 47, 58, *59*, 65, 68, 86–7
  armour maker / *irw ṯryn* 68
  as ethnic stereotype 23–35
  as prize of war 35–3
  as tribute 38–40
arrow maker / *ḥmw ꜥḥꜣ.w* 68
Ashkelon 47, 132 n.6
Asyut: 15–16
Aufrère, Sydney 18–19, 117 n.35, 147 n.3
Ay 29, 126 n.51

Baal *see* Seth / Baal
Battle of Kadesh 2, 62
  Bulletin 54, 58, 95–6, 101
  Poem 54, 58, 95–6, 101
  reliefs 32, *33*, 45, *46*, 58, *59*
Battle of Visby 7–8
Beqaa Valley 8, 37, 74
Beth Shan (Tell el-Hosn) 81–3, 145 n.52
Birch, Samuel 5
Bonaparte, Napoleon 3
Bonnet, Hans 5
bow 2, 14, 17, 21, 38, 42, 43, 80, 97, 103, 115 n.20, 120 n.7

bow maker / *ir.ty pḏt.w* 68
Buhen 68, *69*
Burton, Harry 86

Carter, Howard 86, 147 n.67
Champollion, Jean-François 3, 5
chariot 1, 2, 13, 25, *26*, 27, 29–30, 31–2, 45, *46*, 48, 54, *59*, 65, 66–8, 99, 103, 105, 119 n.2, 124 n.34, 153 n.75
  chariot maker / *ḥmw mrkbt / ḥmw wrrt* 64, 68
  charioteers 22, 27, *30*, *31*, 42–3, 77–8, 140 n.10
  role on the battlefield 1–2, 21, 120 n.4
  chariot driver / *kdn/kḏn* 42–3
  chariot warrior / *znny* 42–3
Collombert, Philippe 94
copper alloy 10, 11, 70, 73, 75, 78, 81–5
Coulon, Laurent 94
Cros, Gaston 11

Dapur *see* Ramesses II, siege of Dapur
Dawson, Doyne 2
Dawson, Timothy 8
dbn n tp / ḏꜣḏꜣ 37, 66, 90–1, 147 n.2
De Backer, Fabrice 7, 12
Deir el-Bahari 17–18, 117 n.35
Deir el-Bersha 16
Dendra panoply 40
Dezsö, Tamás 7, 97
Djehutyhotep 16–17
Dra Abu el-Naga 40

Eannatum 12–13
Ebla (Tell Mardik) 13–14
Euphrates 24

Fortress of Tjaru (*Pꜣ ḫtm n Ṯꜣrw*) 48, 133 n.9

Gebel Barkal 36, 92
Gezer 132 n.6
Girsu (Tello) 11, 12, 14, 15
   copper alloy helmet *11*, 13
Giza 97
group writing 96, 151 n.53

Hanson, Victor Davis 22
Harvey, Stephen 23
Hassan, Selim 97
Hatshepsut 24, 118 n.36, 122 n.20, 23, 123 n.24
Hattusa (Boğazkale) 98
Hattusili III 98
Hauron 97
Healy, Marcus 9
helmet 2, *4*, 5, 6, 7, 22, 25, *26*, 27, 28, 30, 31, 32, 35, 41–3, 45, 47, 49, *50*, *52*, 54, 61, *62*, 65–6, 68, 73, 86, 90–1, 94, 104–5, 135 n.28, 142 n.22
   Aegean 79–80, *81*
   Akkadian 14–15, 138 n.1, 150 n.26
   as ethnic stereotype 23–35
   as prize of war 36–7
   as tribute 39–41
   Sumerian 9–14
Herihor 30
Hittites 22, 28, 32, *33*, 45, 54, 58, 60, 78, 93, 98, 121 n.12, 127 n.62
Horemheb 27, 29–30, 42, 126 n.51, 131 n.104
   mortuary temple *30*, *31*
Horus 68, 97, 100
Howard, Dan 8, 12, 16
Hulit, Thomas 8–9, 40, 41, 147 n.71
Hurrian War *see* Second Syrian War
Hyksos 21, 23–4

Ikram, Salima 8
Intef (TT 386) 19
inw 38–9
Ishtar 98
Isis 97–8
Israelites 47, 132 n.6

Johnson, Raymond 28, 29, 125 n.37
Junge, Friedrich 97

Kāmid el-Lōz *see* Kumidi (Kāmid el-Lōz)

Kamose 21, 119 n.2
Kanakia 83–5
Karnak, temple complex 27–8, 54, 92, 95, 125 n.37, 48, 126 n.59
   'Cour de la Cachette' 47
   Court between Ninth and Tenth Pylon 55
   Eighth Pylon 37, 47, 92
   Great Hypostyle Hall 31–2, 32, 33, 47, 54–5
   Naos of Philippus Arrhidaeus 25, 92
   Second Pylon 28, 29, 42, 131 n.102
   Sixth Pylon 36, 92
   Temple of Khonsu 29–30, 42
   Vestibule of Thutmose III 92
Keegan, John 1
Kendall, Timothy 6, 14, 27, 97, 142 n.25, 150 n.30
Ky-iry *64*, 64–6
Kikkuli 22
King's sandal maker / *ṯbw nsw* 68
khopesh see sickle-shaped sword (*khopesh*)
Krauß, Rolf 6
Kumidi (Kāmid el-Lōz) 8, 74, *76*

Lagash (Tell al-Hiba) 12
Lisht 69, 74–7
Liverani, Mario 35, 38
Libyans 49, *51*, 80
Luxor, temple 28, 29, 42, 54, 61, 125 n.48
   Court of Ramesses II 55, 56, 58, 60, 95
   Pylon 29, 55, 56, 95

Malqata 69–70, 73–4, 141 n.16, 142 n.18
Mari (Tell Hariri) 13
Maryannu 22, 25, 26, 38, 121 n.13, 14, 15
McBride, Angus 9
Medamud 28, 41
Medinet Habu *see* Ramesses III, mortuary temple (Medinet Habu)
Megiddo *see* Thutmose III, battle of Megiddo
Memphis 128 n.76, 140 n.10
   pꜣ ḫpš 64, 67
Merenptah 47, 48, 77, 78
   Victory Stele 47, 132 n.6
Mesehti 15–16
Mitanni 22, 24, 28, 36, 41, 90, 93, 98, 103, 123 n.24

Mödlinger, Marianne 9
Montu 25
mss n ꜥḥꜣ 36–9, 91–3, 94–5, 149 n.19

Naram-Sin 14, 116 n.24
Nebhepetre Mentuhotep II 18–19
neck guard 27, 40
Necropolis of Sheikh Abd el-Qurna 39–41
Nefer-renpet 68
Nubians 28, 29, 49, 131 n.102
Nuzi (Yorghan Tepe) 6, 7, 14, 27, 73, 93, 138 n.1

Odysseus 79–80
Onomasticon Golenischeff 68, 96, 141 n.13

Pap. *Amherst* IX (*Astarte Papyrus*) 58
Pap. *Anastasi* I 65
Pap. *Anastasi* III 47–8, 77, 143 n.37
Pap. Bibliotheque Nationale 202 58
Pap. British Museum EA 74100 80, *81*
Pap. *Chester Beatty* III 56, 96
Pap. *Koller* 99, 153 n.75
Pap. *Lansing* 48
Pap. Louvre AF 6347 100
Pap. Mag Brooklyn 100
Pap. *Mallet* 88–9, 96, 97
Pap. *Raife / Sallier* III 54, 56, 96, 136 n.37
Parrot, André 13
Parkinson, Richard Bruce 80
Pi-Ramesses (Qantir) 77–9, 80, 82, 143 n.39
  area Q I 77–8
  area Q IV 77, 143 n.38
  Pharaoh's garrison 77
Prisse d'Avennes, Émile 3, 5, 58

Qantir *see* Pi-Ramesses (Qantir)
ḳrꜥw see shield-bearer (*ḳrꜥw*)

Radwan, Ali 98
Ramesseum *see* Ramesses II, mortuary temple (Ramesseum)
Ramesses II 32–3, 45–7, 78, 83–5, 88, 92, 95, 101
  body armour 54–61
  mortuary temple (Ramesseum) 54, 55, 57, 58–9, *59*, 62, 96, 127 n.61, 141 n.13

siege of Dapur 2, 58–8, 61–2, 95–6, 101, 138 n.54

Ramesses III 47, 88
  military campaign against Sea Peoples *48*, 49, *50*, 134 n.17,
  mortuary temple (Medinet Habu) 29, 4–9, *50–1*, *52–3*
  undated Asiatic campaigns 52, *53*
  Second Lybian War (Year 11) 49, *51*
  siege of Tunip 52, *53*
  tomb (KV 11) 3, *4*, 54, *61*, *62*, 135 n.27–8
Ramesses IV 88
rbš / lbš 98–100
representations of helmets and armour in private tombs
  Amenmose (TT 42) 39
  Menkheperasoneb (TT 86) 39
  Min (TT 109) 39
  Paimose (A.13) 40–1
  Qenamun (TT 93) 40–1
  Rekhmire (TT 100) 40
  Sennufer (TT 96A) 41
Rimush 14–15

Saqqara 42, 64, 67, 140 n.9
Schneider, Thomas 58
Schofield, Louise 80
Schulman, Alan 6, 43
Sea Peoples 45, 48–9, 50
  Sherden 47, 49, 92, 132 n.2
Second Syrian War 28
Senusret I 74–5
Senusret II 16
Senusret III 16
Seth / Baal 55–8, 62, 138 n.59
Seti I 31–2, *32*, *33*, 57, 78
Shaushka 98
Shasu nomads 24, 122 n.23
Shed 97
shield 1, 13, 14, 17, 27, 42–3, 47, 65, 78, 104, 114 n.12, 144 n.42
shield-bearer / *ḳrꜥw* 42–3
sickle-shaped sword (*khopesh*) 2, 68, 95, 105, 120 n.8, 9
Šilwa-Tešub 73
Skinner, Lucy-Anne 8
Spalinger, Anthony 43, 61, 120 n.4, 132 n.7

Speiser, Ephraim Avigdor 6
Stadelmann, Rainer 97
Suppiluliuma I 28, 98
Susa (Shush) 14, 15, 116 n.24, 117 n.26

Tadu-hepa, daughter of Tushratta 41, 93
Tanis 57, 92
Thebes 19, 27, 28, 29, 39, 40, 41, 43, 48, 54, 58, 88, 100, 124 n.35, 125 n.37, 40
Thordeman, Bengt 7–8
Thutmose I 122 n.23
Thutmose II 24, 122 n.23, 123 n.24
    memorial temple (Qurna) 24
Thutmose III 24, 91, 122 n.20, 23
    Annals of 35–6, 38, 90, 91, 92, 93, 147 n.1, 148 n.4
    battle of Megiddo 35–6, 128 n.72
    Victory Stele from Gebel Barkal 36, 92
Thutmose IV 129 n.82
    tomb (KV 34) 25
    chariot body 25, *26*, 27
Tunip *see* Ramesses III, siege of Tunip
Tushratta 93
Tutankhamun 74–5, 80
    body armour 86–8
    mortuary temple ('Mansion of Nebkheperure at Thebes') 27–9, 42, 124 n.35, 125 n.37

tutiwa / *tutittu* 14, 115 n.21
ṯ3ỉ 55–7, 100–1
ṯryn 55–7, 59–60, 68, 88, 93–8

Umma (Tell Jokha) 12
Ur
    copper alloy helmets *10*, 10–11
    King's Grave (PG 789) 10–11, 12
    Standard of 12–13, 14

Veldmeijer, André 8
Ventzke, Walter 8, 74
Vikentiev, Vladimir 97
Vogel, Carola 19, 117 n.35

Wiese, André 34–5
Wilkinson, John Gardner 5
Winlock, Herbert 17–19
Wise, Terence 9
Wolf, Walther 5
Wolley, Leonard 10

Yadin, Yagael 5, 12, 40
Yenoam 31, *32*, 126 n.56, 132 n.6

Zivie-Coche, Christiane 140 n.9

www.ingramcontent.com/pod-product-compliance
Lightning Source LLC
Chambersburg PA
CBHW052113300426
44116CB00010B/1651